What in the World is a Fecal Transplant?

Written by:
Austin Mardon, Amir Ala'a, Alyssa Wu, Amna
Abu Askar, Angel Xing, Amna Zia, Alexandra
Hauser, Anna Yang, Alexia Di Martino, Alexa
Gee, Amal Rizvi, Ami Patel

Edited by:
Jenny Kang

GM

PRESS

Typeset and Cover Design by Kim Huynh

ISBN 978-1-77369-257-9
Golden Meteorite Press
103 11919 82 St NW
Edmonton, AB T5B 2W3
www.goldenmeteoritepress.com

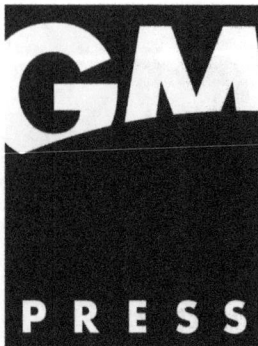

Contents

What is the history behind fecal transplants?

The digestive system is one of the most important systems in the body. It includes vital organs such as your mouth, esophagus, stomach, small intestine, large intestine, rectum, and anus. The pancreas, gallbladder, and liver assist them on their journey delivering food through your body. One of the digestive system's main jobs is to soak up all the nutrients from the food you eat, therefore, making the digestive system a crucial component of your existence according to Cleveland Clinic (2018). Your large intestine is made up of more than just cells. It is also an ecosystem, with trillions of bacteria known as "gut flora" living there. But no worries, most of these bacteria are beneficial. The large intestine and a portion of the small intestine are home to friendly bacteria. The stomach's acidic atmosphere prevents bacteria from growing states (FlexBooks, 2018).

Gut bacteria play a variety of roles in the human body and some examples are:
  • create vitamin B12 and vitamin K
  • limit the spread of harmful bacteria.
  • break down toxins in the large intestine.
  • break down certain food ingredients that are difficult to absorb, such as fiber, starches, and sugars. Bacteria make enzymes that break down carbohydrates found in plant cell walls. Without these bacteria, much of the nutritional value of plant material will be lost. These aid in the digestion of plant foods such as spinach, according to FlexBooks (2018).

Villines (2019) states that beneficial bacteria help the digestive system absorb nutrients and digest food effectively, but certain medical issues and medicines can kill these healthy bacteria. One way to reintroduce efficiency to your system is through a fecal transplant. A fecal transplant is a procedure in which a doctor transplants feces from a healthy donor into another person's gut to restore bacterial equilibrium. Fecal transplants could aid in the treatment of gastrointestinal infections and other illnesses, according to Villines (2019). Now knowing the importance of the digestive system and how a fecal transplant can positively impact an individual's system that may need a proper balance of bacteria, the question of how the condition was discovered, studied, and treated throughout the years must be answered.

 The main records of fecal transplantation date back to fourth century China, where "yellow soup" was applied in instances of serious food contamination and diarrhea. By the sixteenth century, the Chinese had fostered an assortment of dung determined items for gastrointestinal grumblings just as foundational side effects like fever and pain (de Groot et al., 2017). In the meantime, Bedouin bunches were said to have devoured the stools of their camels as a solution for bacterial dysentery, states de Groot et al. (2017). Italian anatomist and specialist Acquapendente (1537–1619) further stretched out this to an idea he authored "transfaunation," the exchange of gastrointestinal substance from a beneficial to a wiped-out creature, which has since been applied broadly in the field of veterinary medicine (de Groot et al., 2017). Strangely, numerous creature species are found to normally rehearse coprophagia, prompting a more prominent variety of microorganisms in their digestive tracts, empowering them to process a more noteworthy number of food sources (de Groot et al., 2017).

The invention of high-throughput DNA sequence analysis, various other -omics technologies, and computational advances allowed the scientific revolution that led to modern studies of indigenous microbial communities, (Khoruts, 2017). It is now clear that the pathogen-centric germ theory of disease, which has dominated medicine since the 19th century, has failed to carefully consider the critical functions of other factors in disease in the physiology of the body; the resident microbiota plays a significant role (Khoruts, 2017). As a result, for decades, doctors have prescribed antibiotics with no regard for the possible long-term consequences of altering the body's microbiota. Khoruts (2017) states that antibiotics have been used in agriculture in far greater quantities, where their advantages, ironically, have nothing to do with pathogens and more to do with the function of the person's indigenous microbiota in growth and energy metabolism. We are now dealing with a wave of multidrug-resistant pathogens as a result. Furthermore, it is assumed that the current epidemics of obesity, autoimmunity, and allergic diseases are linked to population-wide changes in gut microbiota composition (Khoruts, 2017).

Gradually these thoughts started to sprout interest in eighteenth century European doctors. German philosopher Christian Paullini (1643–1712) was quick to diagram the remedial capability of human discharges in his work "Heilsame Junk Apotheke" (in a real sense: healing mud pharmacy), states de Groot et al. (2017). The major disclosure by Antoni van Leeuwenhoek was that his stool contained microorganisms – "God's littlest creatures"– just as perceptions from the Russian zoologist Metchnikoff (1845–1916), established an early framework for the innovative field of microbiota study (de Groot et al., 2017). Enlivened by the reports of life span in Bulgarian ranchers regardless of their helpless day to day environments, Metchnikoff presented aged items in his eating regimen and noted upgrades in his overall wellbeing. He estimated this to be because of a modified equilibrium in colonic microorganisms, with an expansion in lactic corrosive microscopic organisms (actually called "Lactobacillus bulgaricus") ensuring against senescence-speeding up poisons (de Groot et al., 2017). Metchnikoff's microbes caught public attention and were effectively promoted during his lifetime. His idea of expanding the quantity of useful microorganisms in the gut trying to improve human wellbeing, obviously shows a noteworthy utilization of probiotics avant la lettre, states de Groot et al. (2017).

In a comparative light, German bacteriologist and doctor Alfred Nissle confined an Escherichia coli strain that to date bears his name. At first, the microorganism was found defensive against Shigella outgrowth and ensuing gastroenteritis. However, its effect on human wellbeing was subsequently stretched out to incorporate persistent provocative conditions (de Groot et al., 2017).

Before long, gut microorganisms would likewise be discovered significant in recuperating from gastroenteritis. At the point when German fighters of the 'Afrikakorps' were passing on privately contracted looseness of the bowels in the mid 1940s, Nazi researchers were resolved to discover a reason and fix it. Perceptions of better-faring local people who might take in new camel stools upon the primary indications of disease drove them to dissect the dung and disengage Bacillus subtilis, states de Groot et al. (2017). Ensuing refinement and organization of the bacterium settled the illness in many ways. At this point, the various instances of organisms affecting human wellbeing have made way for a more intensive assessment of its applications, and the primary controlled examinations utilizing remedial fecal suspensions have arisen.

Anti-infection agents, since their revelation liberally recommended, have finished a time in which irresistible sicknesses were the most widely recognized reason for death. Nevertheless, anti-toxins likewise accompanied results and antimicrobial opposition. While trying to improve blow-back on commensal organisms, according to de Groot et al. (2017) bacteriologist Stanley Falkow examined fecal material from careful patients prior to beginning them on pre-procedural anti-infection agents. After changing over the stools into pill structure, he recommended their day-by-day admission to half of the gathering during post-careful recuperation, a thought so repulsive at the time it got him terminated when the authoritative load up was discovered. Narrative proof from this examination in the mid 1950s depicts better results in the treatment bunch, however the information of the informal 'Ersatz'10 preliminary was rarely distributed (de Groot et al., 2017).

In the succeeding year, a gathering of Colorado-based specialists driven by Dr. Eiseman played out a trial along a similar reasoning: that rebuilding of a sound gut microbial equilibrium can improve patients' wellbeing. After various treatments had neglected to effectuate recuperation, they treated 4 patients fundamentally sick with pseudomembranous colitis with fecal bowel purges from sound contributors, states de Groot et al. (2017). Results were amazing, with fast and complete recuperation in all subjects. In the accompanying twenty years, 16 additional cases were chosen to go through a similar strategy. That reached a 94% achievement rate, despite the helpless visualization in this treatment, recalcitrant patient gathering demonstrated promising (de Groot et al., 2017).

At this point, the reason for this dangerous condition had been recognized as contamination by the gram-positive anaerobic spore forming bacterium Clostridium difficile, regularly incited by anti-microbial use. As a result of clinical similarities among irresistible and non-irresistible sorts of colitis. Doctors began to estimate whether fecal microbiota transplant (FMT) could likewise be of help in fiery inside sickness and bad tempered entrail disorder (IBD and IBS). The earliest record of FMT for a non-irresistible illness concerns a 45-year-old male having obstinate ulcerative colitis (UC), showing full and enduring clinical recuperation upon "a trade of gut flora" (de Groot et al., 2017). Various ensuing contextual investigations zeroed in on a large number of gastrointestinal complaints. In most cases, patients were found to "[positively] react to manipulation" of their microbiome following the colonic organization of a combination of microflora from unaffected people (de Groot et al., 2017).

By the turn of the century, new bits of knowledge interfacing gut organisms to the pathophysiology of extra-intestinal infections would widen the uses of FMT.

Pseudomembranous colitis was a terrible illness in the 1950s, with a death rate of nearly 75%. The 30-day mortality rate associated with colon surgical excision, which is still the standard model of treatment for antibiotic-refractory severe-complicated C. difficile infection (CDI), is still exactly 50%, states Khoruts (2017). Fischer and colleagues reported a 91% clinical cure of antibiotic-refractory severe-complicated CDI using FMT in 33 consecutive patients in this issue of gut microbes, based on their revised single-center experience, states Khoruts (2017). For clinicians who have directly experienced the remarkable rapid reversal of a C. difficile-triggered toxic megacolon's worsening clinical path after FMT, normal surgical treatment can no longer appear to be an ethically justifiable first choice. In fact, following the Eiseman publication, fecal enemas were quickly introduced in the treatment of pseudomembranous colitis in some centers, only to fade into obscurity just after the introduction of vancomycin (Khoruts, 2017).

More than 700 patients are accounted for to have been treated with FMT for repetitive CDI since Eiseman's experiment (de Groot et al., 2017). Three ongoing randomized controlled preliminaries have revealed fixed paces of 90% or higher overall, the therapy impact is lasting, and protected, with no connected results or recently procured ailments during follow-up, in any event, when acted in weak patient groups (de Groot et al., 2017).

As the weight moved from irresistible to non-transmittable issues, the scope of FMT applications broadened. According to de Groot et al. (2017), ongoing case-reports even incorporate coincidental discoveries of post-FMT reduction in extra-gastrointestinal conditions like numerous sclerosis, Parkinson, idiopathic thrombocytopenic purpura, and persistent fatigue. Another microbiota-related sickness is kwashiorkor (extreme hunger), which Smith and associates concentrated on in 2013 in kids in Malawi. They researched 317 twin sets for 3 years (from birth until 3 years old) (de Groot et al., 2017). During these 3 years, half of the twin sets remained very much supported, however 7% showed concordance for intense lack of healthy sustenance and 43% got harsh treatment for malnutrition. Later on, they relocated fecal microbiota from conflicting sets into sans germ mice. Mice that got gut microbiota from kwashiorkor youngsters showed critical weight reduction, alongside dysregulation of starch and amino corrosive metabolism (de Groot et al., 2017).

Methods striving to handle the opposite finish of the metabolic range, the stoutness pandemic, have neglected to produce acceptable results. Nevertheless, creature and human proof focuses on a potential job for gut microbiome control in restoring energy homeostasis. The accompanying area will momentarily survey how the microbiome has gotten associated with the pathophysiology of metabolic condition and how FMT may help future treatment (de Groot et al., 2017).

Trial proof in creatures associating the intestinal microbiota to the metabolic disorder is largely accessible. Inferable from the formation of sans germ (GF) mice by prof. Jeffrey Gordon and innovative advances in sequencing techniques, researchers had the option to demonstrate causality of microbial association in weight, the executives, and glucose and lipid digestion.

For instance, Bäckhed and partners actuated weight acquire and expanded insulin opposition in GF (C56BL/6) mice upon oral organization of fecal material from their customary partners, despite a concurrent decrease of food intake (de Groot et al., 2017). Scientists credited this to a more compelling starch take-up (resulting in lipolysis prompting increased muscle to fat ratio content) because of preparation of supplements by the microorganisms present. Conventionalization of GF mice likewise enhanced the weight acquired after putting these mice on a Western or high-fat eating regimen (HFD) (de Groot et al., 2017). Next, the beneficiary GF mice of excrement from fat (ob/ob) contributor mice put on altogether more weight contrasted and beneficiaries of defecation from slender benefactor mice according to de Groot et al., (2017).

According to de Groot (2017), Ley et al., (2005) proceeded to show that under comparative dietary conditions, ob/ob mice convey altogether not so much Bacteroidetes but more Firmicutes in their guts contrasted and lean ob/+ and wild-type mice. A new report with

Sprague-Dawley rodents affirmed this microbial mark, where fecal transfers from rats on a control diet re-established an ascent in plasma, unsaturated fat, and glucose narrow mindedness after fructose-prompted metabolic syndrome (de Groot et al., 2017). Strangely, glucose resistance improved altogether through the organization of an anti-toxin blend of ampicillin and neomycin. Ridaura and partners (2013) were quick to relocate human dung into GF (C57BL/6J) mice and affirmed expanded weight and endless supply of fecal material got from a corpulent grown-up contrasted with that of her lean twin.

At last, de Groot et el., (2017) states that a new report by Liou et al. (2013) recommends that the positive metabolic impacts of Roux-en-Y gastric detour (RYGB) medical procedure is halfway because of a modified structure of the gut microbiota. They have performed FMT in GF-mice from patients that had gone through RYGB medical procedure. This technique brought about a huge deficiency of weight and fat mass contrasted with the GF-mice that had gotten gut microbiota from mice that had gone through a trick method. Information approving these impacts in people are energetically anticipated.

In a similar period, further accentuation was put on the part of gut microbiota in metabolic sickness in people, and on human microbiota variety. Of note, Arumugam depicted distinctive human enterotypes and Karlsson showed that investigation of gut microbiota arrangement can anticipate metabolic status, according to de Groot et al. (2017). Curiously, the principal record of abrupt weight acquired following FMT dates from 1983, as a coincidental finding after the goal of intermittent CDI in a grown-up female patient (de Groot et al., 2017). Another such instance of "new-beginning corpulence," was distributed as of late, justifying alert in considering the utilization of corpulent giver material for fecal transplantations.

Concerning the treatment of metabolic condition with FMT, the lone human investigation to date was performed by Vrieze et al., (2012), states de Groot et al., (2017), which recommends that FMT from lean unaffected contributors briefly builds fringe insulin affectability (with a comparative yet not measurably huge pattern toward improved hepatic insulin resistance). In accordance with results acquired from the creature models, these progressions were discovered to be decidedly corresponded with an expansion in the quantity of butyrate-delivering microscopic organisms in the gut. Taking everything into account, past discoveries propose a causal connection between the gut microbiota and metabolic condition (de Groot et al., 2017).

FMT has a long line of descent, aging all the way back to the fourth century where people suffering with severe diarrhea would orally consume what was called "Yellow Soup." This procedure became widely known and started its own timeline of history causing multiple experiments that even lead to studies involving animals, such as mice. The procedure's aim is to create a balanced, diverse microbiome in the gut. Therefore, fecal transplants have been a huge help to the health of many people which we can thank its history for.

Why were fecal transplant treatments developed? Why are other treatments ineffective?

Overview of FMT

Fecal microbiota transplantation (FMT) is commonly used for Clostridium difficile (C. difficile) infections, metabolic syndrome, and inflammatory bowel disease (van Nood et al., 2014). C. difficile is an anaerobic, spore-forming, gram-positive Bacillus that is transmitted through the oral-fecal route (Grigoryan et al., 2020). It is a gastrointestinal infection that affects the natural microbial composition of the intestinal microbiota. FMT allows the recipient's gut microbiota to be replenished with healthy microorganisms transferred from a healthy donor stool sample (Gough et al., 2011). It is important to maintain a healthy microbial composition throughout the body because these microorganisms help break down complex carbohydrates, replenish energy storage facilities within the body and provide protection against pathogen invasion (Sullivan et al., 2001). Other terms for describing the medical process for the infusion of human feces include 'donor feces infusion', 'fecal transplantation, and 'human intestinal microbiota transfer' (van Nood et al., 2014). FMT is a highly successful treatment therapy, which is commonly used when standard treatments have been unsuccessful. FMT was developed to treat a variety of diseases related to gastrointestinal (GI) imbalances with the gut microbiota where standard therapies were deemed to be unsuccessful.

Protocol and ethics for FMT treatments

The typical process for giving FMT to a patient can vary depending on several factors including quantity of donor stool used, recipient preparation, methods for infusion of donor stool, and measurement of outcomes (Gough et al., 2011). Methods for infusing donor stool into the patient include enema, gastroscope tube, or the nasojejunal tube. A normal saline solution is typically used to prepare the stool suspensions before administration to the patient. Following the infusion, patients receive antibiotic treatment to control for side effects and reduce the risk of potential infections (Gough et al., 2011). Microbial feces can be administered through the upper or lower GI tract through the liquid suspension of a stool sample from a healthy donor (Gough et al., 2011). There are many factors to consider when screening for potential donors for FMT. For ease of use in a clinical setting, many donors are usually a close relative of the patient receiving FMT treatment. It is important to consider the exclusion of donors with contractible diseases or those that have an increased risk of transmitting unknown pathogens.

In human microbiome research, there are five main ethical issues including patient consent and respect for autonomy, informing subjects of the research study results, data sharing and privacy, sampling invasiveness and minimization of associated risks, and subject diversity (Daloiso et al., 2015; McGuire et al., 2008). FMT needs to maintain certain safety and ethical standards to safeguard the health of the patient undergoing FMT. The Federal Drug Administration (FDA) has classified the use of human feces as a type of medical drug treatment. This has allowed physicians to perform FMT more readily and expand the availability of this therapy to more patient groups (Donia et al., 2014; Vyas et al., 2015). Donor selection is an ethical issue that has unique challenges associated with recruiting health donors. Donors need to be carefully selected to avoid causing a new disease in the patient. To ease patient concerns, the stool samples should be obtained from a donor for whom the patient has provided informed consent. The licensed health care provider is responsible for conducting the appropriate screening and testing metrics for the donor and the stool sample, before administering FMT to the patient (Daloiso et al., 2015).

Physicians may choose to repeat the treatment on a patient case-by-case basis for patients who were unsuccessful in their treatment or have experienced a relapse in their symptoms. Patients may also be treated with antibiotics, such as vancomycin or metronidazole, which are used to target and relieve symptoms of bacterial infection (Gough et al., 2011).

The problem with colonizing gut microbiota and bacterial resistance
The human GI tract is made up of a complex system of microbes that strengthens human health. The gut microbiota facilitates the process of nutrient synthesis and digestion. Our gut microbiome also

helps our immune system by protecting the body from the invasion of pathogens. The gut microbiome is constantly developing during our childhood and progresses throughout adulthood. When a human infant is born, their immune system is very naïve and immature. This explains why a child is very susceptible to contracting a wide array of diseases as a newborn (Li et al., 2014). The infant's immune system has not had the time to develop its own immunological tolerance to fight the pathogens in our environment. The gut microbiota helps build this immune tolerance by stimulating the production and development of suppressive cytokines, regulatory T cells, transforming growth factor-β (TGF-β) and interleukin-10 (IL-10). These immune molecules help the T helper 1 (Th1) and T helper 2 (Th2) cells form a targeted immune response to the pathogen (Li et al., 2014; Rautava & Isolauri, 2002). When there is an imbalance of these immune molecules, this leads to a poorer active immune response, which contributes to developing immune-related disorders. An insufficient Th1 response leads to the onset of inflammatory bowel disease (IBD); while an insufficient Th2 response leads to the development of allergy (Azad & Kozyrskyj, 2012; Field, 2005).

The intestinal mucosa of the gut contains antigens, food, toxic substances, and organisms that are beneficial to the host (van Nood et al., 2014). The intestinal mucosal immune system is made up of three mucosal lymphoid structures, which are the Peyer's patches, the lamina propria, and the epithelia (see Figure 1). There is a layer of mucus on the surface of the epithelial cells, which acts as a physiological barrier to serve as the host's first line of defense. The second line of defense is the epithelial cells themselves, which acts as the immune surveillance for the gut by communicating and sending signals to the mucosal immune system through the production of cytokines and chemokines. The lamina propria is found in the lower layer of the intestinal epithelial cells, which is a barrier made up of B and T cells. They respond to the lumen of the environment by initiating inflammatory and anti-inflammatory signals to eliminate damage from invading pathogens (Richards et al., 2016).

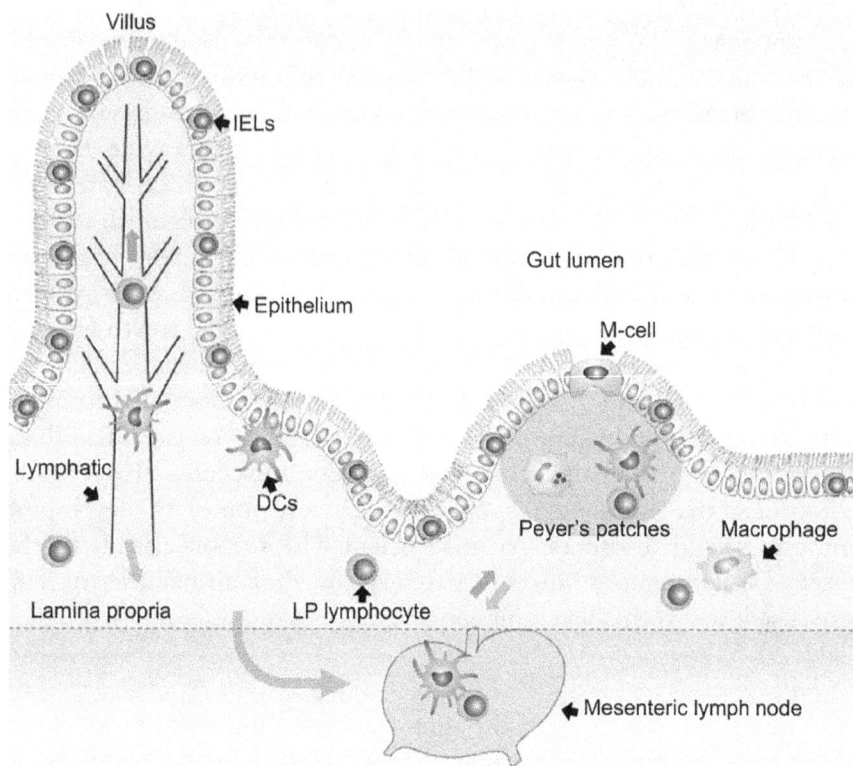

Figure 1. The anatomy of the mucosal immune system (Source: Wu et al., 2014).

Bacterial colonization resistance occurs when the disturbed microbiota has built a tolerance for invading pathogens, which prevents further infections from occurring (van Nood et al., 2014). Bacterial infections characterized by intestinal dysbiosis, such as C. difficile, are not responsive to standard antibiotic therapies. Many bystander effects are associated with using antibiotics, such as the damages incurred on the commensal gut microbiota (Khoruts et al., 2015). Antibiotic-resistant bacteria (ARB) lead to systemic infections that are difficult to treat in a clinical setting. The colonization of ARBs prevents the natural colonization of GI bacterial niches, leading to overall colonization resistance in the gut epithelium. However, antibiotics and chemotherapy treatments have been developed as a way to mitigate these results, but they do impair the natural integrity of the gut microbiota (Bilinski et al., 2017). Two bacterial divisions, known as Firmicutes and Bacteroidetes, are important in colonizing gut bacteria. Metagenomic and biochemical analyses show that these changes affect the metabolic processes (such as nutrient breakdown and energy homeostasis) within the lumen of the gut environment (Turnbaugh et al., 2006).

Many physical barriers may interfere with preventing a patient from undergoing FMT. Many patients have trouble finding a suitable donor. It is imperative that FMT donors are matched with suitable patients to avoid adverse side effects. The healthcare system puts alot of effort into matching donors with recipients to protect their health. According to biological biomarkers, such as commensal gut microflora composition and physical examination, this is a very time-consuming process as laboratory tests need to be conducted to determine whether the donor can be accurately matched with the recipient (Khoruts et al., 2015).

Adverse effects associated with FMT

There are some concerns regarding the safety of FMT. Current research shows that FMT is safe. There have been very few cases reporting serious adverse effects in randomized clinical trials (van Nood et al., 2014). It is common for most patients to experience diarrhea on the first day of infusion. Some patients report symptoms commonly associated with colonoscopy including upper GI tract bleeding, peritonitis, or enteritis (Daloiso et al., 2015; Gough et al., 2011). If there is interference with the bowel passage, vomiting can be a side effect for FMT patients (van Nood et al., 2014). Other side effects include fever and abdominal tenderness. Similar to many other treatment options, these side effects can be adverse following treatment, but many clinical studies have reported that the impacts of these side effects diminish after 2 days (Daloiso et al., 2015).

Long-term risks of FMT therapy have been studied in animal models, as current human data is sparse. Results from these studies show that it is possible for FMT to transfer certain disease phenotypes such as obesity and metabolic disorders. In addition, exposure to antibiotics or cecal content transfer can increase the risk of colitis, heart disease, and colon cancer (Grigoryan et al., 2020).

Why are donor feces needed to treat metabolic disorders?

The gut microbiome is involved in the development of fat mass and altered energy homeostasis (Everard & Cani, 2013). Metabolic disorders are associated with the development of cardiovascular disease and diabetes. In diabetes, insulin is a key molecule that is affected by impaired fasting glucose and glucose tolerance (Eckel et al., 2010). Recent evidence shows that gut microbes are responsible for reducing inflammatory signals that characterize metabolic diseases, leading to gut barrier dysfunctioning (Everard & Cani, 2013). Clinical studies have found that obesity is a major risk factor for patients undergoing FMT therapy, as many obese patients have a lower microbial diversity (van Nood et al., 2014).

Lean donors significantly improve insulin sensitivity in obese patients with metabolic disorders. These types of donors were also found to induce differential microbial modifications in FMT patients, but replenishing their altered microbiome with a slightly increased microbial diversity. In fact, administering stool samples from lean donors into the patients has shown to have an impact in modulating the weight of the patient (Aron-Wisnewsky et al., 2019). This can be beneficial for patients who should gain or lose weight to reach their healthy body mass index (BMI). As mentioned, obese patients can lose weight from received FMT from lean donors. However, in patients with anorexia nervosa, they can gain weight which helps the patient normalize and stabilize their weight over time (de Clercq et al., 2019).

Why are probiotics ineffective for many FMT patients?

Probiotics are commonly used as a therapeutic intervention for IBD, including Crohn's disease and ulcerative colitis. Many drugs have been studied for use as probiotics, however, there are still many adverse side effects as a result of using corticosteroids, aminosalicylates, and immunosuppressants (Verbeke et al., 2014). There is a lack of bacterial strains in probiotics, which makes this therapy incomparable to FMT therapy. A close examination of the chemical composition of probiotics has determined that probiotics are missing essential bile acids, proteins, and bacteriophages, which are critical components of the fecal stool sample used for FMT patients (van Nood et al., 2014).

Vancomycin is a chemical drug that is used to treat infectious diseases caused by bacteria (Ianiro et al., 2018). Vancomycin is a very potent drug that is susceptible to symptoms of toxicity, including nephrotoxicity (kidney toxicity) and ototoxicity (ear toxicity). Pharmacokinetic studies have shown that vancomycin is able to be absorbed in the GI tract, which puts patients with staphylococcal or clostridial diarrhea at risk of developing further disease (Marsot et al., 2012).

Other implications for FMT

Individuals can have an altered microbiome as a result of using antibiotics, which makes these patients more susceptible to various types of infections. To treat bacterial infections, such as C. difficile, antibiotic therapy can cause detrimental GI disruptions in the gut microbiota, which can have a wide variety of lifelong applications for the patient. GI disruptions can lead to enteric disease which can result in mild diarrhea to having life-threatening infections. Once there are imbalances in the GI tract, the risk of recurring infections becomes much higher (Grigoryan et al., 2020).

There are many functional diseases of the GI tract, such as IBS and Crohn's disease, which are currently hypothesized to be very effective in being treated by FMT. IBS is characterized by altering the intestinal microbiota in a way that influences the immunity and motility of cells in the gut microbiome (Daloiso et al., 2015; Pinn et al., 2015). Nonalcoholic steatohepatitis (NASH) is a chronic liver disease in children and adults. The microbial changes in the gut can lead to hepatic cirrhosis (the hardening of the liver) and lesions with the liver (Brunt, 2001). The altered gut microbiota in patients with celiac disease seems to be multifactorial and many symptoms of celiac disease can be treated by FMT (Golfeyz, 2018). Allergies can result from a reduced gut microbial diversity during childhood and adulthood. This is due to impaired mucosal immunoglobulin A (IgA) levels during infancy, which is linked to the development of allergies (Dzidic et al., 2017). Autoimmune disorders, such as rheumatoid arthritis, IBD, and multiple sclerosis (MS), have the potential to be treated by FMT because they are most commonly associated with a change in bowel flora (van Nood et al., 2014). Because the gut microbiome communicates with the nervous system through the gut-brain barrier, neurodevelopmental disorders also can be treated by FMT. The microbes in the gut can communicate with the brain by traveling through the immune system, the vagus nerve, or through the use of short-chain fatty acids (SCFA) to stimulate brain development, function, and behaviour (Kelly et al., 2017).

Conclusion

FMT is a very beneficial therapy for correcting GI imbalances within the gut microbiome. Physicians need to carefully select the appropriate donor(s) for the patient undergoing FMT to ensure that the fecal sample contains the bacteria and microbes that are suitable for the patient while preventing the risk of any new transmission of diseases.

Microbes can colonize in the gut epithelium, which leads to bacterial resistance. This feature is common in pathogens, which renders the therapeutic intervention ineffective. Probiotics are a common treatment used to clear bacterial infections and GI disturbances, however, there are many side effects that are associated with taking these drugs. In addition, clinical studies have shown that probiotics do not contain essential chemical compounds that are naturally occurring in human fecal samples. FMT can be used to treat a wide variety of conditions, including cardiovascular diseases and neurological disorders.

Why is it important to have fecal transplants and what is the impact on individuals?

Fecal transplantation encourages the growth of a healthy gut microflora, which can aid in curing some infections and reducing the severity of gut-related illnesses, by transferring liquid filtrate feces from a healthy donor (Villines, 2019). These transplants can be delivered orally in the form of capsules, through a colonoscopy, nasogastrically, or via an enema. Normally, the gut houses millions of beneficial bacteria that help in efficient digestion and nutrient absorption (Villines, 2019). The gut microbiota acts as a physical barrier against pathogens and plays an important role in mediating our immune system and maintaining intestinal homeostasis (Hooper et al., 2012). When the species diversity and balance of the gut microbiota is disrupted, diarrhea or other gastrointestinal issues can develop (Villines, 2019). The term "dysbiosis" was coined to describe this microbial imbalance, and is not only associated with gastrointestinal conditions such as irritable bowel syndrome, but with metabolic disorders, allergic diseases, autoimmune diseases, and neuropsychiatric disorders as well (Xu et al., 2015). Hence, fecal transplantation can be used to restore the gut microflora and potentially reverse the aforementioned diseases.

In fact, this bacteriotherapy was originally introduced to treat Clostridium difficile colitis, a colon inflammation caused by the Clostridium difficile bacteria (CDC, 2021). Most cases of C. difficile occur in individuals who have been taking antibiotics, which along with eliminating the harmful bacteria, can destroy the good bacteria inhabiting the gut, resulting in dysbiosis and giving harmful bacteria an opportunity to invade and occupy the gut (CDC, 2021). C. difficile has been reported to kill 15,000 people in the United States each year (Villines, 2019). Fecal transplantations, however, have been shown to be highly effective at treating C. difficile associated diseases and prevent its recurrence, which is often seen in patients treated with the standard antibiotic therapy. In a small randomized controlled trial by Youngster et al. (2014), 70% of participants did not present any symptoms following a fecal transplant treatment, and an overall 90% cure rate was observed amongst those who had multiple treatments. The participants have also ranked their overall health status more highly after the fecal transplants (Youngster et al., 2014). The suggested mechanisms by which fecal microbiota transplantation helps against C. difficile infections include introducing a gut bacterial community that competes with C. difficile for a limited amount of nutrients, inhibits its growth and toxigenic activity, modulates metabolites and production of bile acids which can impair the life cycle of C. difficile, or through physiological cross talk between the healthful bacteria and the body's immune system that results in a regulated immune response (Choi & Cho, 2016). Despite the high success rates in treating the recurrence of C. difficile, its practice is limited by concerns of the possible transmission of pathogens, the viability of fresh samples, difficulty in the preparation and administration of the fecal matter, concerns about performing the transplant procedure in a medical office due to odor, difficulty persuading the patients, and the lack of a standard regimen (Choi & Cho, 2016).

While fecal transplantation has been used clinically to treat C. difficile, it may also help with irritable bowel syndrome (IBS) and inflammatory bowel diseases (IBD) (Choi & Cho, 2016; Villines, 2019). Although IBD is not associated with a particular pathogen, it is often characterized by a decrease in butyrate-producing microbes and bacteria from the Bacteroides and Firmicutes phyla, as well as an increase in proinflammatory bacteria like Actinobacteria and Proteobacteria (Frank et al., 2007). It is still unclear whether gut dysbiosis causes IBD or is a consequence of the disease, but there is evidence to suggest that fecal transplantation may be a promising therapy for managing such diseases. The fecal transplants can reintroduce metabolite-producing microbes and restore butyrate concentrations, which helps maintain the integrity of the epithelial lining in the gut. The transplant can also inhibit the differentiation and activation of T-helper cells (Th1 cells), and the production of inflammatory factors and cytokines (The Microbiome, Fecal Microbiota Transplants and Inflammatory Bowel Disease - Mayo Clinic, 2019). However, there appears to be variability in the efficacy of the treatment for IBD (Colman & Rubin, 2014). Hence, fecal transplants have not been approved by the FDA for the treatment of IBD.

Although not applied clinically, restoring gut microbial diversity via fecal transplantation may provide symptomatic relief in patients with Parkinson's disease (PD). Parkinson's disease is the second most common neurodegenerative disease and is characterized by motor deficits including tremors, bradykinesia (slowness of movement), impaired gait, and muscle rigidity (Sampson et al., 2016). Since less than 10% of all Parkinon's disease cases are due to hereditary and genetic factors, Parkinson's is considered a multifactorial disorder (Sampson et al., 2016). Over the years, the aggregation of alpha-synuclein proteins has been found to be the main pathology associated with Parkinson's disease (Sampson et al., 2016). These aggregates result in oligomeric species that accumulate and eventually damage dopaminergic neurons, also referred to as motor neurons, in the substantia nigra pars compacta (SNpc) of the brain (Sampson et al., 2016). As a result, dopamine replacement therapies such as L-DOPA have been widely used to reverse the symptoms of Parkinson's disease. However, the effectiveness of the dopamine modulators are limited by some serious side effects such as dyskinesia or involuntary movements (Jenner, 2008). Neurodegenerative diseases are often studied in the central nervous system (CNS), however, some peripheral influences like the gut may have implications on the onset and progression of neurological disorders. Increasingly, research on the gut-brain axis and the influence the gut microbiota may have on the brain has been heavily investigated. Based on a study performed by Devos et al. (2013), Parkinson's disease patients often exhibit gut inflammation. Specifically, an increase in proinflammatory cytokines including TNF-alpha, IFN-gamma, IL-1beta and IL-6, which are also seen in IBD, were observed in Parkinson's disease patients (Devos et al., 2013). Furthermore, gut microbial compositions were reportedly different in Parkison's disease patients compared to healthy controls. Parkinson's disease patients had lower levels of Clostridium coccoides, Bacteroides fragilis, and hydrogen-producing bacteria, with an increase in Lactobacillus compared to healthy controls (Hasegawa et al., 2015). It is unclear how this dysbiosis arises and whether it is a feature contributing to Parkinson's disease onset and progression. However, physiological alterations associated with Parkinson's disease including reduced gut motility, altered intestinal absorption, or dietary habits may be factors that influence the composition of the gut microflora. It has long been hypothesized that alpha-synuclein aggregates associated with Parkinson's initiates in the gut and enters the brain via the vagus nerve in a prion-like manner. In support of this notion, a study by Svensson et al. (2015) demonstrated that patients who underwent a vagotomy (surgical removal of the vagus nerve) had reduced risks of developing Parkinson's disease. Moreover, alpha-synuclein aggregates seem to appear in the enteric nervous system (ENS), which governs the function of the GI tract, and in the glossopharyngeal and vagus nerves (Braak et al., 2003). Given that Parkinson's disease patients possess an altered gut microbiome, Sampson et al. (2016) examined the effect a PD gut microbiome may have on the brains of healthy rats by transplanting fecal microbiota from PD diagnosed individuals and matched healthy controls into germ free mice. Mice who received PD donor-derived microbiota had an altered short chain fatty acid (SCFA) profile, with reduced acetate concentrations and increased abundance of propionate and butyrate compared to mice who received microbes from healthy controls. To test the effect these alterations have on motor function, the mice completed several motor tasks including beam traversal, pole descent, and nasal adhesive removal. All of which were significantly impaired in mice colonized with PD-donor derived gut microbiota compared to mice with healthy gut microbiota. This indicates that gut microbiota is necessary to promote neuroinflammation and Parkinson's disease symptoms, however, certain microbial metabolites that are seen in PD patients can further exacerbate motor dysfunction (Sampson et al., 2016). The mechanisms by which gut microbiota mediate the pathophysiology of Parkinson's disease are fairly complex, but some studies have demonstrated an active role of the microbiota-produced SCFAs in mediating the activation, maturation and inflammatory capabilities of microglia in the brain. Therefore, fecal transplantation may be a safe and efficacious therapy for the treatments of Parkinson's disease symptoms by restoring gut homeostasis.

Other neurological conditions such as autism may also be implicated with etiologies in the gut. Autism spectrum disorders are characterized as neurological disorders that lead to impaired social interactions and repetitive behavior (Kang et al., 2017). Although autism spectrum disorders (ASD) are poorly understood, gut microbiota has been implicated because many individuals affected by ASD often suffer from gastrointestinal issues such as constipation and diarrhea (Chaidez et al., 2014), which might be in part due to a dysbiotic gut microflora. Those affected by ASD often undertake heavy antibiotics treatments during the first three years of their life, which can disrupt the gut microbiota and allow for pathogenic microbes to colonize and infect the gut (Niehus & Lord, 2006). Previous studies have shown that children with autism have fewer fermentative bacteria and an overall reduced bacterial diversity (Kang et al., 2013). In a longitudinal clinical trial, where 18 children with ASD and with moderate to severe gastrointestinal symptoms participated in a 10-week fecal microbiota transplant (FMT) treatment and an 8-week follow up period after the cessation of the treatment, an 80% reduction in gastrointestinal symptoms after the intervention with enhancements in constipation, diarrhea, and abdominal pain was found (Kang et al., 2017). Moreover, an increase in microbial diversity was observed, specifically the abundance of Prevotella, Bifidobacterium and other taxa, which correlated with improved ASD behavioral symptoms including social skills and daily living skills (Kang et al., 2017). Unlike benefits following a vancomycin therapy, these changes persisted 8 weeks after the FMT treatment. Thus, fecal transplants provide a promising approach to enhance the behavioral symptoms and gastrointestinal symptoms associated with ASD (Kang et al., 2017).

Although many studies have demonstrated beneficial effects following fecal transplant therapies in various disorders, and have suggested that fecal microbiota transplants are safe, especially after the thorough screening of healthy donors, the long term safety of the fecal transplants remains unknown (Villines, 2019). Recently, a fatality resulting from a severe antibiotic resistant infection following a fecal transplant has led the U.S. Food and Drug Administration (FDA) to cease the use of FMT and suspended all clinical trials that use fecal transplants (Villines, 2019). In addition, the use of antibiotics after fecal transplants may result in unwanted complications (Villines, 2019). Hence, more clinical trials evaluating the long-term safety of the FMT treatment are needed. In a study by Meighani et al. (2016), several predictors of fecal transplant failure were identified, including being female, being previously hospitalized, and having a recent surgery before the fecal transplant. Additionally, most studies evaluated the effects FMT had on individuals with a dysbiotic gut microbiota. However, a study by Goloshchapov et al. (2019) monitored the long term effects of FMT on the gut microbiota of healthy individuals. The participants consistently exhibited gut microbial compositions that were similar to that of the donor and a persistence of the effect for about a year. The fecal transplant was well tolerated by the healthy participants with no or mild adverse effects.

In summary, FMTs, although a novel procedure, have been around for a long time. Fecal transplants are becoming of increasing interest as its potential application is growing beyond the treatment of C. difficile infections. When all other treatments fail, this bacteriotherapy may be used to treat conditions including diabetes, insulin resistance, obesity, asthma, eczema, nonalcoholic fatty liver disease, neuropsychiatric and mood disorders such as dementia and depression (Choi & Cho, 2016). By naturally restoring the health and microbial diversity of the gut, gastrointestinal associated illnesses and the person's overall health can be improved. Compared to other therapies such as antibiotic therapies, FMT has been shown to have a high success rate in treating gastrointestinal and colon infections, specifically C. difficile. Although several studies have demonstrated the effectiveness and safety of the FMT procedure, whether it is delivered nasogastrically or via a colonoscopy, many feel anxious about the procedure. However, with careful screening of the donors, transmission of potentially lethal pathogens to a recipient can be avoided.

What is Fecal Microbiota Transplantation?

Introduction

Fecal microbiota transplantation (FMT), also known as fecal transplants, may have a daunting name but is an effective treatment for gastrointestinal diseases with over a thousand years of history. As discussed in previous chapters, fecal transplants were important and are important for the human body when necessary, especially when other treatment methods were unsuccessful. In this chapter, we will explore fecal transplants in detail, discussing how it affects the body and its use in the treatment of various diseases. By taking a closer look at gut microbiota composition, we can better understand why gut dysbiosis, the changing of the gut microbiota composition through different factors like antibiotic use, can negatively affect the human body. While fecal transplants are primarily used to treat gastrointestinal diseases, they can also also be used to treat psychiatric disorders and metabolic disease because the entire body is affected by the gut. That is to say, the range and depth of fecal transplants exceed gut health. Thus, this chapter's section on the uses of fecal transplants to treat diseases will discuss the treatment of Clostridioides difficile infection (CDI), inflammatory bowel disease (IBD), Crohn's disease, inflammatory bowel syndrome (IBS), and psychiatric disorders like Alzheimer's.

What is Fecal Microbiota Transplantation (FMT)?

FMT, also called a stool or a fecal transplant, is the infusion of liquid filtrate feces from a healthy donor into the gut of a recipient for the treatment of a specific disease (Choi & Cho, 2016). Liquid filtrate feces, also called the fecal suspension, are administered into the recipient's gut through several different ways: a nasogastric tube inserted through the nose to reach the stomach, a nasoduodenal tube inserted through the nose to reach the duodenum, colonoscopy, enema, or capsule (Choi

& Cho, 2016). However, there are concerns about donor selection, screening, standardization protocols, long-term safety, and regulatory issues. This will all be explored in chapter six, which will elaborate on the mechanisms of this infusion or engraftment of liquid filtrate feces.

1. Gut Microbiota Composition

Gut microbiota in the intestines exists as a biological barrier against pathogens (Choi & Cho, 2016). The human gut microbiota includes 1000 trillion bacteria, quadrillion viruses, fungi, parasites, and archaea, where the most abundant bacteria phyla are Bacteroidetes and Firmicutes. There are also bacterial species belonging to the phyla Actinobacteria, Fusobacteria, Proteobacteria, Verrucomicrobia, and Cyanobacteria (Choi & Cho, 2016). From birth, each individual would have their own gut microbiota composition based on environmental factors like diet, lifestyle, the use of antibiotics, and hygiene preferences. Given that gut microbiota can be influenced by external factors, it is not fixed. When the gut microbiota composition changes, there can be negative implications on the body. This change of gut microbiota composition is called gut dysbiosis, which can lead to gastrointestinal diseases (Choi & Cho, 2016). Several examples include CDI, IBM, and IBS which will all be explored in the next section.

2. Effect of Antibiotics

Antibiotics have a strong impact on inciting gut dysbiosis (Choi & Cho, 2016). Researchers found that short-term antibiotic use can shift gut microbiota to long-term dysbiotic states (Lange et al., 2016). As well, post-antibiotic dysbiosis can reduce colonization resistance against invading pathogens. By interfering with bacteria reproduction, antibiotic use can affect the size of the gut bacterial community and composition (Choi & Cho, 2016). In the case of CDI, the spores for the bacteria Clostridioides difficile (C. difficile) can germinate and expand. C. difficile was identified in 1978 as the primary cause of pseudomembranous colitis, the severe inflammation of the colon. While antibiotic-associated diarrhea is one of the most common adverse side effects related to antibiotic use, CDI can result in toxic megacolon and potentially death (Mullish & Williams, 2018). The use of antibiotics can result in osmotic diarrhea caused by the loss of gut bacteria that can absorb short-chain fatty acids, but also the colonization and overgrowth of C. difficile. Since antibiotics kill bacteria, they can cause the loss of gut microbial communities that protect against gut infection, enabling the germination and growth of C. difficile (Mullish & Williams, 2018). Additionally, individuals greater than 65 years on acid-suppressant medications, immunosuppressants, and were hospitalized for greater than seven days or admitted to a room where the previous patient had CDI are at higher risk for CDI (Mullish & Williams, 2018). Thus, antibiotics can negatively impact the gut, leading to a potential proliferation of C. difficile which would cause CDI.

3. Gut Microbiota and the Gut-Brain Axis

Additionally, gut microbiota maintain intestinal homeostasis and modulate the host's immune system (Choi & Cho, 2016). This is because of the bidirectional interaction between gut microbiota and the gut-brain axis (Carabotti et al., 2015). The gut-brain axis is the bidirectional communication between the central and enteric nervous system, which links the brain's emotional and cognitive centers with the intestines' peripheral functions. The evidence of microbiota-gut-brain axis interaction is from the association of dysbiosis with central nervous disorders, like anxiety-depressive behaviours, with functional gastrointestinal disorders. One main example is IBS. Carabotti et al. (2015) found that the disruption in the gut-brain axis results in changes in intestinal motility and secretion, causing visceral hypersensitivity, and leading to cellular alterations of the entero-endocrine and immune systems. Simply put, there is a collateral dysregulation of the gut-brain axis and gut microbiota in IBS. Carabotti and the research team found that the principal mechanism of how microbiota may interact with the gut-brain axis is the modulation of the intestinal barrier. Gut microbiota interact with the central nervous system by regulating brain chemistry and influencing neuro-endocrine systems. Therefore, the gut microbiota does not solely affect gastrointestinal processes as there is a bidirectional relationship between gut microbiota and the gut-brain axis.

Consequently, gut dysbiosis can also lead to metabolic diseases, autoimmune diseases, allergic disorders, and neuropsychiatric disorders (Choi & Cho, 2016). Focusing on metabolic health, the gut microbiota has the functional capacity to induce or relieve metabolic syndrome (Marotz & Zarrinpar, 2016). Although obesity and related metabolic disorders were primarily linked to human genetics and lifestyle habits, the gut microbiota is becoming more recognized as playing a key role. First, Marotz and Zarrinpar found that certain metagenomic patterns of the gut microbiota are responsive to weight changes, which suggests that changes in the gut's microbiota correlate with the human's metabolic phenotype. Second, despite the individuality of gut microbiota based on environmental factors, researchers found that there is a specific set of bacterial functional gene profiles that are present in all healthy individuals (Marotz & Zarrinpar, 2016). In short, scientists believe that gut dysbiosis can lead to metabolic syndrome because specific gut bacteria and their metabolites affect the host's metabolism and feeding behaviour (Lee et al., 2019).

Given that gut microbiota relates to metabolic diseases, scientists are currently conducting clinical studies to test treatments that can alter gut microbiota (Marotz & Zarrinpar, 2016). Probiotics, non-pathogenic organisms that benefit the host, are one method. However, there were no studies that reported an alteration in fecal microbiota composition. Therefore, Marotz and Zarrinpar turned to FMT, which is proven to have significant changes in fecal microbiota composition. Furthermore, FMT has a successful history in the treatment of gastrointestinal diseases like CDI. Unfortunately, at the time of Marotz and Zarrinpar's study, FMT was only a proposed method of treatment. In 2019, scientists were still testing FMT through controlled clinical studies to see if it can effectively treat metabolic syndromes (Lee et al., 2019). Ideally, FMT would be a new therapeutic option for obesity and associated metabolic disorders, but more studies are necessary.

All in all, FMT, also known as fecal transplants, is an established treatment for gastrointestinal diseases like CDI, and a potential treatment for psychiatric and metabolic disorders. As there is more research on how the gut microbiota affects the entire body, there could be more uses for FMT. Gut microbiota includes a plethora of organisms that facilitate a variety of functions: acting as a barrier against pathogens and modulating intestinal homeostasis, the host's immune system, and metabolism. Growing research on the gut-brain axis finds that the gut microbiota and the gut-brain axis have a bidirectional relationship. Unfortunately, gut microbiota can be negatively affected by antibiotics, resulting in gut dysbiosis which leaves the body more susceptible to infection from C. difficile. Fortunately, this can be remedied through FMT. The next section will focus on how FMT is used to treat diseases.

How is Fecal Microbiota Transplantation Used to Treat Diseases?

1. Clostridium difficile infection (CDI)

CDI is primarily caused by gut dysbiosis, the imbalance of gut microbiota (Konturek et al., 2016). Common symptoms include diarrhea, peripheral leukocytosis, acute renal failure, hypotension and pseudomembranous colitis. To treat CDI, healthy gut microbiota must be restored. However, Konturek and the team noted that from a clinical point of view, the increasing prevalence of antibiotic resistance inhibits the treatment of CDI. Antibiotics are commonly used to treat infections, but repeated antibiotic therapy negatively impacts gut microbiota diversity. Indeed, antibiotics can successfully treat infections by eliminating bacteria, but it also increases the risk for C. difficile overgrowth and CDI recurrences. If an individual had two or more episodes of CDI, their risk for subsequent CDI recurrence exceeds 60% if they are treated with antimicrobial therapy (Youngster et al., 2014). Fortunately, FMT can restore healthy microbiota, which would eliminate the infection in a rapid and lasting manner (Konturek et al., 2016). In fact, patients treated with FMT exhibit a 90% cure rate. Moreover, Konturek et al. (2016) found that the cure rate for FMT in patients with recurrent and/or severe colitis from CDI was 94%. The most common side effect was abdominal discomfort, but there were no allergic reactions or fever. Konturek and the team also observed a normalizing for C-reactive protein and fecal calprotectin three weeks after FMT, concluding that FMT therapy also has strong anti-inflammatory effects.

FMT can be effectively applied in a variety of ways, such as using a frozen inoculum from unrelated donors, nasogastric tube administration, and colonoscopy (Youngster et al., 2014). Youngster et al. (2014) noted that practical and aesthetic barriers hamper the widespread use of FMT, despite its proven efficacy for treating CDI. Moreover, the recruitment and screening of donors is a lengthy and costly process, which prevents the immediate use of FMT in serious, acute situations. Thus, Youngster and the team thought a collection of prescreened frozen donor stools could make the treatment more accessible and available. They sought to study the clinical outcomes of FMT for refractory or relapsing CDI in patients treated with a frozen suspension from unrelated donors by upper and lower GI routes. 20% of patients reported mild abdominal discomfort and bloating and one child treated colonoscopically had a fever of 38.8 degrees celsius that resolved spontaneously. Despite these side effects, the overall cure rate was 90% at eight weeks. From this, Youngster and the team concluded that the infusion of unrelated frozen donor stools is effective. As well, nasogastric tube administration is a viable route of administration, which would be more accessible for the elderly and debilitated patients who may not tolerate a colonoscopy. However, it is also necessary to note that this study was conducted on a randomized but small group of 20 people.

In 2016, researchers also conducted a study on 201 patients between 48 and 85 years of age who underwent FMT from December 2012 through May 2014, examining their patient factors, diseases factors, and transplant factors to determine the risk factors for FMT failure (Meighani et al., 2016). For the study, scientists defined failure of treatment as the irresolution of diarrhea in patients who were treated with one or more FMTs within 90 days. Meighani et al. (2016) found that the overall failure rate was 12.4% and more common among females, those who were previously hospitalized, and had surgery before FMT. Ultimately, Meighani and the research team concluded that FMT can replace antibiotics for the treatment of recurrent CDIs. Thus, FMT is a key treatment for CDI and recurrent CDIs.

2. Inflammatory Bowel Disease (IBD)

Inflammatory bowel disease is a chronic autoimmune GI disorder that results in chronic inflammation in the gastrointestinal tract, and it is known that gut microbiota is associated with the development and maintenance of IBD (Weingarden & Vaughn, 2017). Weingarden and Vaughn (2017) noted that one of the clearest human genetic associations with IBD is the nucleotide-binding oligomerization domain-containing protein 2 (NOD2). The risk of colitis from NOD2 mutations is often from the subsequent dysbiosis, which can be improved or worsened by changing the gut microbiota. Given that NOD2 plays a role in the clearance of intracellular pathogens through autophagy, a mutation of NOD2 would inhibit this function and result in failures to clear potentially pathogenic microbes. Since FMT is a successful therapy for CDI, it is a potential treatment method for IBD because it would manipulate gut microbiota. Despite attempts with probiotics and antibiotics to manipulate gut microbiota to treat IBD, as explored in the previous chapters, they both have major drawbacks. FMT is a more viable option as Weingarden and Vaughn say that it is a more robust method of manipulating the gut microbiota. By transferring processed feces from a healthy donor to the recipient's GI tract, it increases the diversity of fecal bacterial populations and has long-term effects proven by its historical success in treating CDI (Weingarden & Vaughn, 2017). Therefore, there is growing research into the use of FMT to treat IBD.

FMT is also becoming increasingly popular in the treatment of pediatric IBD (Wang et al., 2016). Pediatric IBD phenotype may have different pathophysiology from adult-onset IBD as the early age of onset would make the cumulative burden of medications, nutritional impairment, and surgery greater than the later age of onset for adults (Wang et al., 2016). Since the pediatric chronic disease would overlap with growth, bone accretion, and psychosocial development, the disease can have more significant long-term impacts. On the other hand, the shorter latency of the disease can allow the pediatric gut microbiota to be more malleable than a fully formed adult, so the immature immune system of children can be more easily influenced by FMT. Wang and the team (2016) found in a systematic review that well-designed, pediatric randomized controlled trials are necessary to further investigate the use of FMT as a therapeutic option for IBD.

Gut microbiota plays an important role in a subtype of IBD: Crohn's disease (Gutin et al., 2019). Considering how patients with Crohn's disease have significant gut dysbiosis, depicted as a decreased diversity and bacterial load, there is an increasing interest in the use of microbial restoration therapies, like FMT, to treat Crohn's disease. For the study, Gutin and the team administered surveys, blood and stool samples prior to and post undergoing FMT to patients aged 18-70 recruited from the University of California San Francisco between July 2015 and October 2016. All the patients underwent a single FMT administration via colonoscopy, and they were contacted at three milestones to assess adverse side effects.

While 30% of patients achieved the primary outcome of improvement one month after FMT, there was no significant improvement in objective measures of inflammation (Gutin et al., 2019). In addition, the study was prematurely stopped because of flare-ups in two patients within days of undergoing FMT. For responders with lower gut microbiota diversity, FMT may provide symptomatic improvement. FMT did increase the relative abundance of healthy bacteria from the donors and reduced bacteria associated with CDI. In the end, Gutin and the team concluded that more data is necessary to determine the predictors of Crohn's disease patients who may experience flare-ups after FMT. Furthermore, the early termination of the study due to adverse effects indicates the necessity to assess the efficacy and safety of FMT in Crohn's disease, in order to determine if microbial restoration therapy with FMT is truly a beneficial treatment option.

3. Neuropsychiatric disorders

While FMT was commonly used to treat gastrointestinal issues, it is also an emerging treatment for neuropsychiatric and neurodegenerative disorders. (Choi & Cho, 2016). With the growing research on the gut-brain axis and the relationship with gut microbiota, there is more knowledge about the relationship between intestinal dysbiosis and changes in mood, behaviour, and cognition (Chinna Meyyappan et al., 2020). Gut microbiota can be influenced by genetics, diet, metabolism, age, geography, antibiotic treatment, psychotropics, and stress. Provided that the gut microbiota is critical for the regulation of brain development and function, the interaction of the gut with the environment, as diet and stress, is a risk factor for psychiatric disorders. In addition, the physiological manifestations of psychological distress as abdominal pain and headaches implicate a connection between the gut and the brain. Currently, there is research on FMT's application in treating psychological disorders, as well as variations of FMT like Microbial Ecosystem Therapeutics-2 (MET-2) to treat Generalized Anxiety Disorder (GAD) and Major Depressive Disorders (MDD) (Chinna Meyyappan et al., 2020). MET-2 uses gut bacteria from healthy donors' stool samples that are later purified and lab-grown before lyophilization, a process to remove the water before it is provided to patients for oral ingestion (Chinna Meyyappan et al., 2020). The research on FMT and MET-2 treatments is an ongoing process.

The use for FMT-related therapies in psychiatric disorders is most commonly studied for MDD and anxiety disorders are the two most common groups of psychiatric disorders (Chinna Meyyappan et al., 2020). Although there are existing treatments, the side effects and stigmatization prove a need for more options. FMT would change the gut microbiota, thereby affecting the gut-brain axis, and potentially improving symptoms of MDD and anxiety disorders. Chinna Meyyappan and the research team conducted a review on 28 studies evaluating the effect of FMT on psychiatric and physical symptoms, finding that the most frequently studied symptoms were from those with irritable bowel syndrome, chronic stress, and depression. Ultimately, the findings demonstrate that FMT does exhibit transmissible properties for the treatment of psychiatric symptoms in both mice and humans. This supports the idea that treatments targeting the gut microbiota can potentially benefit the treatment of MDD and anxiety disorders. However, the mechanism for how the gut microbiota affects the nervous system leading to the symptoms is not yet fully understood. The majority of hypotheses believe that gut microbiota affects serotonin production, immune response, and metabolism, which would subsequently affect depression and mood. Moreover, the gut microbiota can influence the central nervous system by interacting with the vagus nerve. 80% of the vagus nerve is composed of afferent nerve fibres which are affected by the metabolites of the microbiota (Chinna Meyyappan et al., 2020). Additionally, the vagus nerve is affected by long and short-chain fatty acids. Thus, FMT can be a viable treatment for psychiatric symptoms.

Scientists are studying the relationship between gut microbiota and Alzheimer's disease (Jiang et al., 2017). Alzheimer's disease is a neurodegenerative disorder associated with impaired cognition, and it is a common form of dementia. Since the gut-brain axis and the gut microbiota have a bidirectional connection, the effect of the gut on the brain is also explored in Alzheimer's disease. From studies performed on germ-free animals and in animals exposed to pathogenic microbial infections, antibiotics, probiotics, or FMT, Jiang et al. (2017) found that there is a role for gut microbiota in host cognition for Alzheimer's disease-related pathogenesis. Furthermore, gut dysbiosis can cause inflammation related to Alzheimer's disease's pathology. That is to say, the gut microbiota may play a crucial role in the development of Alzheimer's disease. It is necessary to conduct more research into the underlying mechanisms of the gut-brain axis and the gut microbiota to provide more insights into new therapeutic strategies for neurodegenerative disorders.

Conclusion

FMT is becoming an increasingly significant treatment option beyond solely CDI. With increasing knowledge of the gut-brain axis' interaction with the gut microbiota, the potential for FMT expands beyond the gastrointestinal tract. Despite the aesthetic repulsion of infiltrating liquid feces from a healthy donor into the gut of the recipient, it is proven to have high efficacy rates over other treatment methods like probiotics and antibiotics. As well, unlike antibiotics, it does not have the potential negative effect on gut microbiota that can worsen gut dysbiosis.

This chapter focused on two main topics: understanding FMT and an overview of how FMT is used to treat various disorders. Research on FMT, gut microbiota, and the gut-brain axis is a work in progress. As explored in this chapter, there is still much to uncover. More studies must be conducted on larger sample sizes, and more needs to be understood about the mechanisms behind the gut-brain axis and gut microbiota interaction. Nevertheless, the research points towards an optimistic direction for the budding use of FMT therapies.

How are fecal transplants performed today?

As discussed in previous chapters, a fecal transplant, also known as a fecal microbiota transplant (FMT) or bacteriotherapy, is commonly used for the purpose of treating gastrointestinal infections (Villines, 2019). In this procedure, feces from a healthy donor is transferred to a recipient's gastrointestinal tract in order to balance good and bad bacteria in the large intestine and restore their gut health (Healthwise, n.d.). FMTs are most commonly used for the treatment of Clostridium difficile infection (CDI) and there has been growing interest in their use for non-CDI diseases, such as inflammatory bowel disease (IBD) and irritable bowel syndrome (IBS), albeit, with limited or inconclusive evidence on its treatment efficacy (Merenstein et al., 2014).

The administration of FMTs may vary across clinicians as no singular standardized protocol has been universally adopted (Merenstein et al., 2014). This chapter will provide an in-depth overview on the administration of FMT in this day and age, specifically discussing the following elements:

• Recipient Eligibility Criteria — Who can receive an FMT?

• Fecal Transplant Donors
– Donor Selection
– Donor Screening

• Stool Donation and Laboratory Processing

• FMT Administration Procedure
– Upper Gastrointestinal FMT: Nasogastric, nasoduodenal and nasojejunal tubes
– Upper Gastrointestinal FMT: Oral FMT pills

– Lower Gastrointestinal FMT: Enemas
– Lower Gastrointestinal FMT: Colonoscopies
– Best FMT protocol

Recipient Eligibility Criteria — Who can receive an FMT?

Before an FMT is performed, clinicians use indications to determine which patients are offered the FMT treatment (Mullish et al., 2018). It is recommended that the initial treatment of CDI should not involve FMTs, rather, FMT should be considered for patients suffering from recurrent or relapsing CDI (Mullish et al., 2018). Specifically, these patients have had:

• at least three recurrences of mild-to-moderate CDI, where standard vancomycin therapy or alternative antibiotic therapy (e.g., rifaximin, nitazoxanide, etc.) has been ineffective for 6-8 weeks;

• at least two recurrences of severe CDI, where hospitalization is necessary and patient is at risk of developing significant morbidity (Bakken et al., 2011).

FMTs can also be offered to patients who have not necessarily had recurrent CDI, but have had:

• moderate CDIs and are unresponsive to standard vancomycin therapy for at least a week;

• severe or sudden onset of CDI and are unresponsive to standard vancomycin therapy after 48 hours (Bakken et al., 2011).

Regarding co-morbidities and FMT, it is also good clinical practice to offer FMTs with caution to patients with CDI who have decompensated chronic liver disease or are immunocompromised, as well as avoid administering FMTs in patients with anaphylactic food allergies (Mullish et al., 2018). There may also be cases where clinicians will not follow the aforementioned indications. This is because the severity and progression of the patient's CDI should be the key indicator for clinicians to determine whether to proceed with an FMT early on or in a compromised individual (Bakken et al., 2011).

1. Fecal Transplant Donors

To prepare for an FMT procedure, a potential fecal donor needs to be carefully selected and screened for the CDI patient (Merenstein et al., 2014).

Donor Selection

There is general consensus that those unrelated to the CDI patient (i.e., universal donors) or their close contacts (i.e., family members, close relatives, spouses, etc.) can be appropriate candidates (Mullish et al., 2018). CDI patients can access fecal transplant donations from stool banks/pharmaceutical companies that sell donated stool, or choose to receive donated stool from a family member or friend should they not want to purchase it (Jørgensen et al., 2017; O'Neill, 2019). The British Society of Gastroenterology (BSG) and Healthcare Infection Society (HIS) guidelines recommend that an unrelated, healthy donor from a centralized stook bank is used, however there is considerable debate on the preferable source of the donated stool as there are advantages and disadvantages of using either type of donor (Bakken et al., 2011; Bibbò et al., 2020; Mullish et al., 2018). Donors that are intimate contacts (i.e., significant others), may reduce the risk of transferring infectious agents as they share environmental risk factors with the recipients, though there is a greater chance that they too are a C. difficile carrier (Bakken et al., 2011). On the other hand, donors who are first-degree maternal-line relatives may theoretically have the greatest number of intestinal microbial species in common with the recipient. Therefore, FMTs from intimate contacts or relatives may be tolerated better by patients due to the adaptive immune elements in the mucosal immune system (Bakken et al., 2011). However, rigorously screened donors that are unrelated to the recipient may also be advantageous — they are more readily available, which eases the execution of the FMT procedure and may be useful for FMT treatment of genetic diseases such as IBD (Bakken et al., 2011; Kelly et al., 2015). According to the available evidence, both related and anonymous fecal donors are clinically effective in resolving recurrent CDIs and the choice to use one over the other is usually driven by personal needs (Cammarota et al., 2017).

Donor Screening

Potential donors need to undergo two levels of screening — a preliminary interview on medical history and laboratory testing of blood and stool, where a number of factors are evaluated as listed in Table 1 (Bibbò et al., 2020) and Table 2 (Bibbò et al., 2020). Though not an exhaustive list, these factors highlighted in Table 1 and Table 2 represent the most frequently inspected factors in donors at leading FMT centers (Bibbò et al., 2020). The preliminary interview is structured as a questionnaire which aims to gather information on the donor's infectious risk factors, gastrointestinal co-morbidities and microbiota profile (Bibbò et al., 2020). The laboratory testing of blood and stool is then conducted to minimize the risk of transmitting communicable disease via fecal matter (Bibbò et al., 2020). Fecal donors who are sexually intimate partners of the patient may undergo an abbreviated version of this screening process as they most likely share bodily fluids and exposure to infectious agents, where their donated stool will not significantly increase risk of adverse events for the patient (Bakken et al., 2011). This may be particularly useful in FMT procedures that are time-dependent and are not in a position to wait for test results (Bakken et al., 2011). Although there may be situations or conditions where the administration of FMTs is high-risk, considerations regarding the severity of the patient's illness should be prioritized over these contraindications (Bakken et al., 2011). This means that the decision to overlook a certain screening protocol and perform the FMT may override the risk of transmission of a communicable disease from the donor to recipient, if a fecal donor is not found in a timely manner and the patient's clinical deterioration is time-sensitive (Bakken et al., 2011). Due to the grave ramifications of FMTs altering the intestinal microbiome, some researchers are advocating for a more extensive screening process to exclude donors with chronic medical conditions such as atopy, chronic fatigue and obesity (Merenstein et al., 2014). Depending on new research findings, it is quite possible that the list of exclusion criteria for fecal donors will become more stringent in the future (Nicco et al., 2020). Once the donor tests negative for risk factors and is deemed free of chronic diseases in the screening process, they will be allowed to move forward with their stool donation (Bibbò et al., 2020).

Table 1

Components of the donor's medical history that are assessed in the preliminary interview

Preliminary Interview – Medical History
Drugs that can alter gut microbiota
Use in the last three months of: • Antimicrobial drugs • Immunosuppressant agents • Chemotherapy Daily use for over three months: • Proton pump inhibitors

Disorders potentially associated with the disruption of gut microbiota
• Personal history of chronic gastrointestinal disease, including functional gastrointestinal disorders; inflammatory bowel disease; celiac disease; other chronic gastroenterological diseases or recent abnormal gastrointestinal symptoms (e.g., diarrhea, hematochezia, etc.) • Personal history of cancer, including gastrointestinal cancers or polyposis syndrome, and first-degree family history of premature colon cancer • Personal history of systemic autoimmune disorders • Obesity (body mass index > 30) and/or metabolic syndrome/diabetes • Personal history of neurological/neurodegenerative disorders • Personal history of psychiatric/neurodevelopmental conditions
Known history or risk behaviors for infectious disease
• History of HIV, hepatitis B or C viruses, syphilis, human T-lymphotropic virus I and II • Current systemic infection • Use of illegal drugs • High-risk sexual behavior • Previous tissue/organ transplant • Recent hospitalization or discharge from long-term care facilities • High-risk travel • Needle stick accident in the last six months • Body tattoo, piercing, earring, acupuncture in the last six months • Enteric pathogen infection in the last two months • Acute gastroenteritis with or without confirmatory test in the last two months • History of vaccination with a live attenuated virus in the last two months

Table 2

Components of the donor's blood and stool that are assessed via laboratory testing

Blood testing
• Complete blood cell count • Liver enzyme (Aminotransferases) • Bilirubin • Creatinine • C-reactive protein • Serology for Hepatitis virus (HAV, HBV, HCV, HEV) and Human immunodeficiency virus (HIV)
Stool testing
• Clostridium difficile • Giardia lamblia, Cryptosporidium spp, Isospora and Microsporidia • Protozoa and helminths and parasites (including Blastocystis hominis and Dientamoeba fragilis) • Antibiotic-resistant bacteria • Common enteric pathogens, including Salmonella, Shigella, Campylobacter, shiga toxin-producing Escherichia coli, Yersinia, and Vibrio cholerae • Norovirus, rotavirus, adenovirus • Helicobacter pylori fecal antigen

Note. From "Fecal Microbiota Transplantation: Screening and Selection to Choose the Optimal Donor," by S. Bibbò, C. R. Settanni, S. Porcari, E. Bocchino, G. Ianiro, G. Cammarota, & A. Gasbarrini, 2020, *Journal of clinical medicine, 9*(6), 1757 (https://doi.org/10.3390/jcm9061757). CC BY 4.0.

2. Stool Donation and Laboratory Processing

Once the donor and their stool has been adequately screened, they can collect their stool at home and transport it to the FMT production center or donate a stool sample on-site, after which it can be processed in a laboratory setting (Nicco et al., 2020). Laboratory processing of the donor stool sample allows for its preservation and consistent suspension for clinical use (Jørgensen et al., 2017). Stool banks or centers that process donated stool, may prepare frozen (i.e., stored for later use) or fresh FMT (i.e., for immediate clinical use) from recurring donors (Mullish et al., 2018). Therefore, the donor must be able to donate on multiple occasions over a locally-defined donation period (Mullish et al., 2018). Though the exact protocols vary from center to center, donors must also undergo repeat screening (in the form of the aforementioned health questionnaires or laboratory testing) over this period of time, as well as on donation day (Bibbò et al., 2020; Mullish et al., 2018). This is done in order to exclude potential donors who may have had a recent onset of a harmful event that makes their donated stool ineligible for a safe FMT (Bibbò et al., 2020; Mullish et al., 2018).

Standard guidelines on the collection of donor stool do not currently exist (Mullish et al., 2018). General principles of FMT preparation include using sterile collection devices, adopting a hygienic procedure for collection and providing donors clear how-to-at-home instructions to limit the risk of cross-contamination (Mullish et al., 2018). To prepare the stool sample for FMT, usually, stool from a single donor is used per FMT and is liquefied by adding a diluent (Mullish et al., 2018). The volume of diluent used to produce this stool emulsion varies between studies, although common practice follows an approximate stool-diluent ratio of 1:5 (Mullish et al., 2018). The liquification of the donor stool via diluent occurs through resuspension in water, milk or a non-bacteriostatic solution, either by directly mixing the ingredients in a beaker or homogenizing stool and fluid in a sterile bench-top blender (Merenstein et al., 2014). The stool slurry is also filtered through laboratory sieves to eliminate any large chunks of particulate matter or undigested food (Nicco et al., 2020). To minimize time-related degradation or alteration, this processing of the donor stool sample should ideally take place within six hours of defecation (Mullish et al., 2018). Once mixing is complete, the stool emulsion is aliquoted into cryotolerant pots and immediately stored at -80° C — for the preparation of frozen FMTs, an appropriate cryoprotective substance (usually a glycerol solution) is also added to the slurry in order to improve bacterial viability upon freezing (Mullish et al., 2018; Nicco et al., 2020).

3. FMT Administration Procedure

The donor stool sample prepared for FMTs may be administered fresh (i.e., immediately after processing) or from a stored frozen sample that can be thawed when applied to the patient (Mullish et al., 2018). Early FMTs were performed with fresh donor stool samples, however they can prove to be impractical for large scale use (Bibbò et al., 2020). Using fresh donor stool sample may complicate or stall the FMT procedure — donations from donors may be delayed, variations may exist in the quality of the stool, and the donor may have to complete frequent screening before the FMT can be performed on the anticipating patient (Jørgensen et al., 2017). On the other hand, frozen donor stool samples require minimal preparation immediately prior to the administration of an FMT, as those preparation steps are added to the aforementioned laboratory stool processing stage (Jørgensen et al., 2017). Regardless, research has shown that fresh and freeze-thawed stool sample suspension are equally effective and measure similarly in the frequency of adverse events (Jørgensen et al., 2017).

As discussed in previous chapters, FMTs can be administered in multiple different ways, through the upper or lower gastrointestinal tract. These include:
  • via the nose (by means of nasogastric, nasoduodenal, or nasojejunal tubes)
  • via the mouth (i.e., the oral ingestion of pills comprised of fecal matter)
  • via an enema
  • via a colonoscopy (Kelly et al. 2015; Mullish et al., 2018).

Upper Gastrointestinal FMT: Nasogastric, Nasoduodenal and Nasojejunal Tubes

Nasogastric (NG), nasoduodenal (ND) and nasojejunal (NJ) feeding tubes are medical devices used to provide nutrition in individuals who are unable to obtain it orally, also referred to as "feeding tubes" (Whitlock, 2020). The tubes differ with regards to their final destination in the gastrointestinal tract. NG tubes end in the stomach, whereas ND and NJ tubes extend further down — ND tubes end up in the first portion of the small intestine, i.e., the duodenum and NJ tubes end up the second portion of the small intestine, i.e., the jejunum (Feeding Tube Awareness Foundation, n.d.). There has been variability in the practise of administering FMT via NG, ND and NJ tubes (Kassam, 2018). Conventionally, this is an endoscopic procedure consisting of a thin, flexible feeding tube carrying donor stool, being inserted into the patient's nostrils, which then extends down the throat and eventually the stomach (Boston Children's Hospital, n.d.). The donor stool sample is drawn into the nasoenteric tubes via syringes and infused slowly into the patient (Goldenberg et al., 2018). At the same time, the patient is monitored for signs of coughing or vomiting, in which case the FMT is aborted (Goldenberg et al., 2018). Post fecal infusion, patients are to be positioned upright at 45° to a 90° angle for four hours to minimize the risk of aspiration or regurgitation (Cammarota et al., 2017; Kassam, 2018). To reduce procedure-related risk, clinicians may also use radiography or fluoroscopy prior to the FMT treatment in order to determine the optimal placement of the nasoenteric tube (Kassam, 2018).

Upper Gastrointestinal FMT: Oral FMT Pills

Though the delivery of FMT via oral pills encapsulating fecal material is more convenient and less invasive than traditional FMT routes, it is currently a developing field of study (Cammarota et al., 2017; Kao et al., 2017). FMT pills, also known as "poop pills" contain fresh, frozen or lyophilized (i.e., freeze-dried) fecal material, all of which have demonstrated comparable clinical efficacy and safety (Kao et al., 2017). The fecal material contained in these pills is the same fecal material delivered in enema, colonoscopy, or nasoenteric tube FMTs (O'Connor, 2018). Much like any other pill, they are ingested orally and are lined by a casing that disintegrates in the gut due to the acids found in the stomach and colon (O'Connor, 2018). This prevents the development of an aftertaste or a pungent smell in the moth (O'Connor, 2018). In order for these pills to be effective, it is important that the donor stool sample enclosed inside is safe and that the pill is designed optimally to disintegrate at the right time so it may reach the correct area of the gastrointestinal system (O'Connor, 2018). Though overall FMT via oral ingestion of pills seems promising for the treatment of recurrent CDI, further research is required to explore its optimal formulation and dose (Mullish et al., 2018).

Lower Gastrointestinal FMT: Enemas

In an FMT via enema, the donor stool sample is directly introduced into the lower part of the large intestine (Seladi-Schulman, 2019). In order to ease the delivery of the FMT to the intestine, the patient may be positioned to lie on their side while their lower body is elevated (Seladi-Schulman, 2019). Then, a lubricated enema tip is gently inserted into the rectum, through which the donor stool sample (that is contained in an enema bag) is suspended (Seladi-Schulman, 2019). Patients are then to hold the infused fecal material for a minimum of 30 minutes and lie in a supine position to reduce the urge to defecate (Cammarota et al., 2017). The procedure for enemas is very similar to colonoscopies, though enemas are less invasive and expensive (Seladi-Schulman, 2019). However, enemas may need be administered more than once to be equally as effective as colonoscopies or nasoentric FMTs, as there is potential for the donor stool sample to not reach the colon in the first try (Boston Children's Hospital, n.d.; Jørgensen et al., 2017).

Lower Gastrointestinal FMT: Colonoscopies

Much like enemas, FMT via colonoscopy also introduces the donor stool sample directly to the large intestine, however, enema targets the distal colon, whereas a colonoscopy targets the proximal colon or terminal ileum and/or cecum (Ramai et al., 2021). In this method a thin hollow tube attached to a camera is inserted into the colon. Then, a catheter-tipped syringe injects the donor stool sample into the cecum (or the most proximal aspect of the intestine that can be safely reached) via the biopsy channel of the colonoscope (Boston Children's Hospital, n.d.; Goldenberg et al., 2018; Kassam, 2018). It is recommended that the donor stool sample is deposited in the right colon, however in the case of severe colitis, it can be deposited in the left colon for the patient's safety (Cammarota et al., 2017). The patient is then asked to resist the urge to defecate and retain the stool for as long as possible (Allegretti et al., 2014). Prior to the procedure, the patient may also be sedated in order to assist the patient in retaining the stool after the procedure (Allegretti et al., 2014). To prolong the retention of donor stool samples in the patient's intestines, clinicians may have patients take loperamide before or after the FMT colonoscopy (Goldenberg et al., 2018; Kump et al., 2014). However this is optional as FMTs with or without loperamide have reported to be clinically successful and its use may actually exacerbate FMT colonoscopy side effects, such as bloating, abdominal pain, nausea or emesis. (Kassam, 2018; Kump et al., 2014).

Best FMT Protocol?

Currently, literature on the best route of fecal instillation is inconclusive (Merenstein et al., 2014). Despite procedural differences, FMT via colonoscopies, nasoenteric tubes or enemeas for the treatment of CDI have all been reported to be successful and also happen to be the most frequently used methods (Bakken et al, 2011; Jørgensen et al., 2017). Each FMT route has its own procedure-related adverse events that clinicians can consider on a case-by-case basis (Kassam, 2018). Administration of an FMT with nasoenteric tubes is associated with a risk of aspiration, though this risk can be minimized by infusion of the donor stool sample beyond the pyloric sphincter (Kassam, 2018). There are also some unresolved issues pertaining to oral FMT pills, specifically that their capsules are large and patients are required to ingest 30 capsules in one day (Mullish et al., 2018). This task can be very challenging to commit to, even more so for elderly patients who may already carry an increased pill burden (Mullish et al., 2018). It is recommended that upper gastrointestinal FMT delivery is avoided or used with caution if the patient has an ileus or a swallowing disorder (Kassam, 2018; Mullish et al., 2018). Colonoscopies also have risk of procedure-related adverse events in rare cases, such as perforation, bleeding, sedation-related aspiration or cardiopulmonary events (Kassam, 2018). At the same time, colonoscopies have added utility compared to other FMT routes since they allow for the diagnosis of other gastrointestinal disorders or cancers during the FMT procedure (Kassam, 2018). On the other hand, enemas are low-risk for the most part and not as invasive, therefore they may be more appropriate in patients who are elderly or have multiple comorbidities (Kassam, 2018). It is also recommended that enemas are the preferred source of FMT delivery for individuals who are critically ill or when lower gastrointestinal FMT via colonoscopy is not possible (Cammarota et al., 2017; Mullish et al., 2018). However, enemas may not be as effective in patients who have poor sphincter tone (Kassam, 2018). Ultimately, healthcare professionals should use their discretion to determine the most ideal FMT method for their patient based on their comfort level and risk factors (Kassam, 2018).

What science is involved in fecal transplants (studying and performing)?

Introduction

Fecal transplants, although sounding fake, are backed by many years of scientific advancements. From links between the gut biome and health to the actual practice of transplanting, fecal transplants are heavily supported by science. The development of fecal transplants relied heavily on the gut-brain axis and the gut microbiome, which serve as the main concepts behind them. Studies have shown many areas where the two interact, and the gut's health can play a prominent role in overall health. The efficacy of fecal transplants has been questioned through research across many diseases and patients. Through a combined look at the science behind the development of fecal transplants and the research on the efficacy of this treatment, a better understanding of the authenticity of fecal transplants and their results will be obtained.

Development

Gut-Brain Axis

The primary scientific basis for fecal transplants comes from the gut-brain axis (GBA), which is the bidirectional link between the central nervous system and the enteric nervous system (Rege & Graham, 2017). It allows for communication via indirect and direct pathways between cognitive and emotional functions and intestinal functions. The GBA allows for complex communication between the endocrine, immune, and autonomic nervous system (ANS). Within the ANS, the GBA primarily functions to combine the sympathetic (responds to stress, increasing arousal) and parasympathetic (opposite to sympathetic, decreasing arousal), which include the afferent and efferent signals sent between the intestines and the brain (Waxenbaum, Reddy, & Varacallow, 2020; Rege & Graham, 2017).

With endocrine communication, the GBA communicates with the hypothalamic-pituitary-adrenal axis to maintain adaptive responses to stress, including activation of memory and emotional centers in the limbic system. Neuro-immuno-endocrine mediators from the GBA influence intestinal function and the cells of the gastrointestinal tract are also influenced by the gut microbiome (Rege & Graham, 2017). This concept where both the GBA and the gastrointestinal tract affecting the intestine is where the scientific basis behind fecal transplants comes from; that there is bidirectional communication between the microbiota and GBA through signalling from the gut-microbiota to the brain and vice versa through neural, immune, endocrine, and humoral links (Carabotti et al., 2015).

Gut Microbiome

As mentioned, the gut microbiome can influence the gastrointestinal tract cells, but the microbiome also plays a role in the overall health of a person. The human microbiome is composed of archaea, bacteria, viruses, and eukaryotic microbes which reside inside the body. They significantly impact the body, from influencing metabolic processes to protecting against pathogens and informing the immune system (Shreiner, Kao, & Young, 2015). Technological advancements have advanced the study of the gut microbiome for performing culture-independent analyses, and the metagenomic analysis by sequencing the microbial DNA has allowed for the assessment of the genetic potential of the population. The accumulation of data on the metagenomics of the microbiome by the European Metagenomics of the Human Intestinal Tract and the NIH-funded Human Microbiome Project has allowed for the creation of a collection of a complete set of genes for most bacteria that would be found in the human gut (Shreiner, Kao, & Young, 2015). Over 750 000 genes were found during this expansion, 30x more than found in the human genome. Less than 300 000 were shared between 50% of individuals, illustrating the size and variability of the bacteria found in the human gut microbiome. During this research, a healthy adult microbiome was identified. The healthy adult microbiome included over 1000 different species of bacteria, belonging to few phyla, the most dominant of which being Bacteroidetes and Firmicutes (Shreiner, Kao, & Young, 2015). Generally, of the bacteria found to be in the human gut, 75% were Bacteroidetes and Firmicutes, which are highly sensitive to changes in their environment (Rege & Graham, 2017). As such, changes in the environment, health, and more can influence these gut bacteria. Since they are very dominant within healthy microbiomes, they can have a considerable influence on the overall health of the microbiome.

One's DNA initially determines the microbiome, and it continues to develop as they are exposed to microbiota as infants in the birth canal and through breast milk. The microorganisms to which the infant is exposed depend on those available from the mother. As they grow, the environment they live in can have either a positive or negative influence on the biome (Harvard School of Public Health, 2020). Other influences on the gut microbiome can come from taking probiotics, as they introduce more robust bacteria to the microbiome, and diet, as high fibre foods break down into short-chain fatty acids, which undergo fermentation, lowering the pH of the gut, and changing which bacteria can live there as a result (Harvard School of Public Health, 2020).

GBA and Microbiome Interaction

Through different studies on the gut-brain axis, specifically one that studied germ-free animals and the effects of antibiotics, probiotics, and fecal transplants on them, many specific pathways have been found to show the interaction between the gut microbiome and gut-brain axis (Rege & Graham, 2017). For example, one of these interactions is with the vagus nerve, where neurons carry feedback from the intestinal end to the brain stem, engaging the hypothalamus and limbic system, affecting the regulation of emotions, and make projections from the limbic system (caused by stress) to the gut influencing its activity. Another interaction is neuroendocrine signalling. The bacteria products in the gut can essentially stimulate enteroendocrine cells, producing different neuropeptides (i.e. peptide YY, neuropeptide Y, cholecystokinin, substance P, et cetera), which enter the bloodstream and influence the nervous system (Rege & Graham, 2017). In particular, peptide YY decreases appetite and makes people feel full and slows the movement of food through the digestive tract. Neuropeptide Y plays a role in coping with stress, and substance P functions as a neurotransmitter and modulator of pain perception (Society for Endocrinology, 2018; Sah & Geracioti, 2012; Graefe, & Mohiuddin, 2020).

Other interactions are with Tryptophan metabolism since most serotonin is produced by enterochromaffin cells located in the gut. Tryptophan metabolism is done before the production of serotonin, and since that occurs in the gut, the gut microbiome also plays a role in it, interfering with the metabolism (Rege & Graham, 2017). Since it is included in the gut-brain axis, the immune system is a large area of GBA and microbiome interaction as gut-associated lymphoid tissue makes up a large portion of the body's immune system. Interactions like altered intestinal permeability and the production of microbial metabolites have also been discovered, the intestinal permeability referring to chronic stress altering intestinal permeability and causing inflammation which has been linked to psychiatric disorders, and the metabolites referring to the bacteria producing multiple neurotransmitters, such as gamma-aminobutyric acid (GABA; the primary inhibitory neurotransmitter), serotonin, and dopamine, which all have significant impacts on the brain's function.

These many different gut-brain axis and gut microbiome interactions and their wide-spanning effects on human health are the scientific basis for fecal transplants. As shown through the interactions, the changing of the microbiome to a more unhealthy one can have significant impacts on the nervous system, among others, causing and contributing to many debilitating disorders.

Efficacy
Many different agencies have completed research on fecal transplants and their efficacy. This research has crossed many various diseases that fecal transplants are supposed to improve, and people exposed to many different environments, cultures, and more. Efficacy can also be assessed through personal anecdotes; however, they may not tell the whole story of the treatment and do not maintain the same level of certainty and reliability as comes with scientifically conducted research.

Cancer
One of the diseases that has seen research on the effectiveness of fecal transplants is cancer. Some patients with cancer have types of cancer that do not respond to immunotherapy drugs. As such, doctors have used fecal transplants to adjust the composition of their gut microbiomes. It has affected their response to types of immunotherapy that were previously not working (National Cancer Institute, 2021). One study had patients with metastatic melanoma, who did not respond to treatment with anti-PD-1 immunotherapy, receive a fecal transplant. They performed a phase 1 clinical trial to test the safety and feasibility of fecal transplants and reintroduced anti-PD-1 immunotherapy in the patients (Baruch et al., 2021). There were clinical responses observed in three of ten patients, with two exhibiting a partial response and a single complete response. This

study shows that treatment with fecal transplants can be associated with favourable changes in what can enter immune cells and gene expression profiles in the gut and tumour microenvironment (Baruch et al., 2021).]

Autism Spectrum Disorder

Autism spectrum disorder is one disorder that has been attempted to treat with fecal transplants. The basis for autism being affected by the gut microbiome comes from a hypothesis that the use of antimicrobial products can impair and alter the gut microbiome, allowing for organisms that can produce neurotoxins into the body. The neurotoxin might be related to the development of autism, but there is little empirical evidence supporting this hypothesis, other than that symptoms of autism tend to be reported around the exact times as they begin using antimicrobial products (Zane & Holehan, 2018). While links between the cause of autism and the gut microbiome are not certain, there has still been research conducted on fecal transplants and autism. In 2017, a study was completed where fecal transplantation was combined with microbial transfer therapy and administered for ten weeks to children with ASD. The result was an 80% reduction in gastrointestinal problems and a steady improvement in core symptoms of ASD (Kang et al., 2017). They found that the gut microbiome diversity, including microbes that had the potential to be beneficial significantly increased after this treatment. Two years later, in 2019, they followed up with the study participants and found that the initial improvements from the therapy remained two years after stopping the treatment and that the diversity of the bacteria in the microbiome was higher for ASD participants. There was also a suggestion from the results that the recipients did not completely retain the donated microbiome, but some features of it, such as the overall diversity (Kang et al., 2019). This study shows promising results, and while there is no evidence that an unhealthy gut microbiome causes ASD, fecal transplants can be used to improve the symptoms of autism spectrum disorder.

Gastrointestinal Disorders

Perhaps the first thought for the use of fecal transplants is for gastrointestinal disorders. From disorders like irritable bowel syndrome (IBS), inflammatory bowel disease (IBD), and Crohn's disease, many studies have been conducted on the effects of fecal transplantation. It has been known that dysbiosis, a low bacterial diversity in the gut microbiome, can contribute to the symptoms and pathophysiology of IBS and other gastrointestinal diseases. For IBS, a double-blind placebo-controlled study published in 2019 where 165 patients were either transplanted with placebo (30 grams of a fecal sample of their own), 30 grams or 60 grams of a fecal sample obtained from a donor. After the treatment, it was found that the fecal transplant patients that experienced a response was similar for both the 30-gram and 60-gram sample (a difference of around 13%), and it increased between two weeks to one month, then maintained after three months (El-Salhy et al., 2019). The final percentage of patients who received the donor sample was 76.9% for the 30-gram sample and 89.1% for the 60-gram sample. It is also important to note that all subjects who received the donor sample received samples from the same donor, and thus the same microbiome was introduced to their gut microbiome. In contrast, those who received the placebo of their own sample started with 49.1% of patients having a response at two weeks and decreased to 23.6% at three months (El-Salhy et al., 2019). This study demonstrated that fecal transplants are an effective treatment for IBS, improving symptoms, fatigue, and increasing the quality of life.

As far as Crohn's disease, another study published in 2019 looked at whether single-doses of fecal transplants can improve clinical and endoscopic outcomes in Crohn's patients. They had ten patients undergo a fecal transplant and were evaluated for clinical response and microbiome profile one month later (Gutin et al., 2019). Three of the patients responded to the transplant, and two had significant adverse effects that required the escalation of their therapy. As far as the gut microbiome, at one month post-transplant, the bacterial communities in the patients' microbiomes increased in the relative abundance of the bacteria common to the donor gut microbiome. They concluded that there was a modest effect with a single dose of fecal transplantation for the treatment of Crohn's disease and the potential for harm (Gutin et al., 2019). Another study that looked at fecal transplants and Crohn's disease patients used a randomized, single-blind, sham-controlled pilot trial of fecal transplants in adults with Crohn's disease. There were 17 patients, where eight received fecal transplants and nine received a sham (placebo) transplantation (Sokol, 2020). Of the patients who received a placebo, the remission rate at 10 and 24 weeks was 4/9 patients and 3/9 patients, and of those who received a fecal transplant, the remission rate was 7/8 at ten weeks and 4/8 at 24 weeks. It was noted that the absence of donor microbiota engraftment was associated with the flare of Crohn's disease. While the primary endpoint of the study was not reached, the study functioned as a pilot and showed that higher colonization by the donor's microbiome was associated with the maintenance of remission of Crohn's disease, but the results must be confirmed in more extensive studies (Sokol, 2020).

Psychiatric Disorders

Fecal transplants have also been studied as a treatment option for psychiatric disorders like depression, bipolar, anorexia and alcoholism. A systematic review of 21 studies on the effect of fecal transplants on psychiatric disorders was published in 2020. This meta-analysis of preclinical, preclinical with human donors, and clinical studies, found that the included studies suggested that fecal transplants can affect symptoms of psychiatric disorders (Meyyappan et al., 2020). This was shown both in forward and reverse of the transplantation, where healthy samples were transplanted into ill recipients, improving their symptoms, and ill samples were transplanted into healthy recipients, worsening their symptoms. This transmissible property was tested in mice and supported by multiple studies demonstrating the transference of symptoms using mouse models of various psychological disorders (depression, anxiety, et cetera). All clinical studies found improvement in symptoms using fecal transplantation from healthy donors (Meyyappan et al., 2020). However, the symptom alleviation and improvement was inconsistent with the length in which it lasted. Some studies (including preclinical) showed that the alleviation was indefinite, while others experienced transient effects; generally, the benefits lasted three to six months (Meyyappan et al., 2020).

Conclusion

There is a lot of rich science supporting the use of fecal transplants. From the initial concepts of the gut-brain axis to the but microbiome and their interactions, there has been a lot of research and studies that support the use of some microbiota transfer treatment. The efficacy of fecal transplants continues to be uncertain, however, many studies are being released that lend to its use in treating disorders from gastrointestinal to the autism spectrum.

What questions are we still asking about fecal transplants?

Introduction

Surveys show that patient perceptions of fecal microbiota transplantation (FMT) are saturated with the idea that FMT is a natural and safe treatment which is somehow separate from conventional medicine, providing an often highly-sought alternative to traditional therapies (Grigoryan et al., 2020). This favourable impression of FMT is often based on sources such as social media and personal anecdotes, and exists alongside the rising interest in the gut microbiome and its interactions with human health, which has thoroughly pervaded mainstream media in recent years (Ekekezie et al., 2020). This combination of factors has stimulated an intense interest in FMT among the general public and, along with the highly accessible nature of donor stool, has paved the way for do-it-yourself (DIY) adaptations of FMT, leading to an increase in unregulated use of this procedure outside of clinical supervision (Grigoryan et al., 2020). Both public interest in FMT and unregulated DIY attempts have quickly outpaced the research and regulatory efforts of the medical and scientific communities. The reality, despite public enthusiasm for the procedure, is that there remains significant unanswered questions regarding FMT, including what its mechanism of action is, what its long-term side effects are, how protocols regarding the screening and transplantation of donor stool should be standardized, and whether or not FMT may be applicable for indications other than C. difficile infection (CDI). This chapter will discuss the current status of research in each of these areas, as well as the specific questions pertaining to each area that remain.

Mechanism of action

Although FMT has been shown to be a highly effective therapy for recurrent CDI, and shows promise for certain non-CDI indications, its mechanisms of efficacy currently remain poorly understood. The research in this area has identified a number of contributing factors that appear to be significant, but there are still significant gaps in our understanding, and more research is needed.

Broadly speaking, it has been established that FMT rapidly restores a gut microbiota that has been disrupted by recurrent antibiotic therapy, causing it to resemble pre-morbid composition and diversity (Smillie et al., 2018). Proof-of-concept studies have inferred the central role of this restoration in the efficacy of FMT by demonstrating that either a healthy donor-derived and defined community of commensal bacteria or fractionated spores from ethanol-shocked donor stool can have a similar efficacy in treating CDI, compared to conventional FMT (Martinez-Gili et al., 2020). However, this picture is complicated by studies that have demonstrated that sterile filtered donor stool can also cause sustained remission from CDI, raising the possibility that it is soluble factors, not intact bacteria, that are the key mediators for the efficacy of FMT (Ott et al., 2016).

Additionally, non-bacterial gut microbiota components have been another key area of focus with regards to attempts to determine the mechanism of efficacy of FMT. Several studies have described an association between the stool donor virome or mycobiome with the efficacy of FMT for CDI, with both undergoing rapid changes in the stool of CDI patients who are successfully treated with FMT (Martinez-Gili et al., 2020). However, the significance of these changes with respect to FMT's mechanism of efficacy remains unclear, and is complicated by a number of facts. For instance, given the established relationship between antimicrobial treatment and Candida overgrowth in the gut, any observed changes in gut mycobiome profiles following FMT may only represent proxies of gut bacterial alterations (Martinez-Gili et al., 2020). As such, the precise nature of the contribution of bacteriophages and fungi to the mechanism of action of FMT remains unclear at this time.

A final significant area of focus among researchers attempting to determine the mechanism of efficacy of FMT concerns the impact of FMT on gut microbial metabolites. A number of different metabolites have received attention as potential contributing factors. Various bile acids, for instance, have been shown to have varied effects on the ability of C. difficile to undergo germination or vegetative growth in vitro (Sorg & Sonenshein, 2008). More specifically, the conjugated primary bile taurocholic acid (TCA) promotes spore germination of C. difficile, whereas secondary bile acids inhibit its vegetative growth (Theriot et al., 2015). In mammals, the conversion of primary bile acids to secondary bile acids occurs within the distal gut by multiple enzymes produced by the gut microbiota The first step is mediated by hydrolase and the second step is undertaken by 7-α-hydroxylase. Studies have demonstrated that the activity of these bile-metabolizing enzymes is protective against CDI in rodents (Studer et al., 2016). As such, a hypothesis regarding the mechanism of action of FMT is that patients with recurrent CDI, which is often accompanied by the antibiotic-mediated destruction of their gut microbiota, are deficient in these bile-metabolizing enzymes. This leads to the enrichment of TCA, which promotes C. difficile germination, as well as the loss of DCA, which facilitates the vegetative growth of C. difficile, thus perpetuating the disease. It follows that a potential mechanism of action of FMT is that the treatment restores the bacteria that produce bile-metabolizing enzymes, reversing the abnormal bile acid environment in the distal gut. This hypothesis is supported by the observation that the stool of patients with CDI is enriched in TCA, whereas secondary bile acids dominate in post-FMT stool (Seekatz et al., 2018). However, more research is needed to determine the significance of bile acid metabolism relative to the other potential factors involved.

Another group of metabolites that has received significant attention is short chain fatty acids (SCFA). In studies involving rodents, higher SCFA levels were associated with protection against C. difficile growth while antibiotics reduced SCFA levels in stool (Theriot et al., 2015). Meanwhile, human studies have shown that levels of various SCFAs within stool are very low in patients with CDI, but are restored to levels that are comparable to healthy donors following successful FMT (Seekatz et al., 2018). That being said, given the established link between antibiotic use, dietary intake, and SCFA production, it cannot be determined from observational studies alone whether the post-FMT increase in stool SCFA levels is a reflection of changes in dietary intake, recovery after antibiotic discontinuation, or the action of FMT (Martinez-Gili et al., 2020). Consequently, the question of the mechanism of efficacy of FMT remains unanswered at this time, although numerous promising avenues of investigation do exist.

Long-term side effects

Currently, information on the long-term effects of FMT in humans remains sparse, as multiple questions remain regarding the immunologic, metabolic, and physiological responses to FMT, as well as the long-term safety implications of altering one's microbiota composition. One factor that has contributed to this scarcity of data is the lack of long-term follow-up periods in most published studies regarding FMT (Woodworth et al., 2019). The current literature identifies long-term prospective cohort studies as a necessity before the potential risks of FMT in humans can be properly and comprehensively assessed (Woodworth et al., 2019).

Despite the limited data currently available on this topic, there is some evidence that indicates that unknown long-term side effects should be a concern. For starters, multiple animal studies have shown that FMT is capable of transferring disease phenotypes such as obesity and metabolic disorders (Grigoryan et al., 2020). Others supplement the hypothesis that FMT can transfer disease phenotypes by demonstrating the impact of other gut microbiome alterations such as cecal content transfer or antibiotic exposure on disease phenotype. For instance, a study has shown that the transfer of antibiotic-perturbed microbiota can increase risk for colitis in mice (Schulfer et al., 2017). More generally speaking, various gut bacteria have also been linked to increased risk for numerous diseases. One study found that the human colonic bacterium enterotoxigenic Bacteroides fragilis may promote colon tumour formation by activating T helper type 17 T cell responses (Wu et al., 2009). Another study has demonstrated that metabolism of dietary L-carnitine, abundant in red meat, by intestinal microbiota accelerates atherosclerosis in mice (Koeth et al., 2013). Furthermore, genomic analysis has identified an association between colorectal carcinoma and Fusobacterium (Kostic et al., 2011).

A number of human case studies further supplement this collection of evidence supporting the existence of unknown long-term side effects associated with FMT. For instance, one well-known case study has documented a human patient who developed new-onset obesity after receiving stool from a healthy but overweight donor (Alang & Kelly, 2015). This case of transferability of obesity with no known biochemical marker for which one can screen donor stool further serves to raise concerns regarding the potential phenotypes that may escape detection during screening processes and subsequently be transmitted to patients undergoing FMT. Additionally, a number of case reports have described a variety of new diagnoses that were temporally associated with the administration of FMT (Woodworth et al., 2019). Although not a substitute for long-term prospective cohort studies in humans, this evidence does suggest that the existence of currently unknown long-term side effects of FMT should be a concern, and that further research in this area is necessary.

The use of FMT in young adults and children, which is ongoing, is of particular concern, as these are the patients who are at the greatest risk of being harmed by long-term side effects. Furthermore, studies have demonstrated that the gut microbiome in early life is strongly susceptible to perturbations, and may also be a driving force in immune development (Bokulich et al., 2016). This means that youth undergoing FMT may be at risk of microbiome alterations that influence their risk of disease later in adulthood (Fujimura et al., 2016).

Protocol standardization

In addition to uncertainty regarding FMT's mechanism of action and long-term effects, there is also little agreement with respect to how the treatment protocol, including donor stool screening and transplantation, should be standardized. There is currently a wide variability in FMT approaches in the published literature, and no universally validated and adopted protocol for the procedure (Grigoryan et al., 2020). In addition, the regulatory future of FMT remains unclear as unregulated, DIY attempts conducted outside of any clinical supervision remain widespread. In this respect, there is a necessity for collaboration among the government, industry, and academia in order to develop and maintain patient-centred regulatory approaches.

One key area of protocol standardization in which many questions remain is donor selection. Little is currently known about the qualities that make someone an effective and safe stool donor for FMT and how to select for these donors, beyond an understanding of the need to screen for certain transmissible pathogens. With respect to stool donor screening, most centres that conduct FMT require donors to undergo stool testing for C. difficile, salmonella, and parasites, and blood testing for HIV, hepatitis A, B, and C (Owens et al., 2013). However, there is, as of yet, no universally validated and adopted screening protocol. Furthermore, current screening protocols may be insufficient, as they do not account for the possibility of gut microbiome perturbation through mechanisms other than known infectious agents (Grigoryan et al., 2020). For instance, the microbiome features of what constitutes a good stool donor remain unknown and in need of study using metagenomics (Borody et al., 2013). The multitude of questions that remain regarding how to select effective and safe stool donors is further complicated by the uncertainty in related areas such as how one can select for an ideal FMT recipient, whether a good stool donor for CDI would be a good donor for other indications as well, or if donors must be matched to recipients in a manner akin to blood and organ transfusions.

Furthermore, there is a high degree of uncertainty surrounding the standardization of transplantation protocol. Once a safe and effective donor is matched with an appropriate recipient, what is the optimal way in which FMT should be performed? There are a number of questions that still need to be answered in this area. For starters, the optimal route of administration for FMT remains unclear. Routes used include nasogastric/nasojejunal tube, endoscopy, oral capsules, retention enema, sigmoidoscopy, colonoscopy, and capsules (Gulati et al., 2020). While it is acknowledged that the safety, efficacy, and cost of FMT depend in part on the route of administration used, the best route for FMT has yet to be established, with extensive studies still being needed in order to understand the interplay of route adopted, donor type, physical nature of sample (fresh or frozen), patient compliance, and cost effectiveness in order to design an approach that is risk-free, convenient, and cost-effective (Gulati et al., 2020). It is also debated whether the induction of an unedited donor sample is required for successful FMT, and whether the transplant material could be whole flora extract or cultured (Grigoryan et al., 2020).

The true determinants of FMT outcomes also remain unclear. This has important implications for protocol standardization, as knowing whether a select consortium of bacteria or the degree of donor microbiome engraftment are the true determinants would inform decision-making regarding how the transplantation should be conducted. Finally, the ideal processing method that should be used, as well as the best size and frequency of the FMT dose, have not been well established (Mullish et al., 2018). With respect to the processing method, it is important to consider whether stool processing steps are necessary in order to preserve metabolites, pH, viruses, or anaerobic non-spore-forming bacteria (Woodworth et al., 2019). These are all questions that still need to be answered, and the current deficiency in evidence and consensus in these areas serve as significant barriers to the development of a standardized FMT protocol.

Applications outside of C. Difficile Infection (CDI)

An additional question that remains unanswered is whether FMT's established success in treating CDI can translate into success for other indications. Although FMT use in the United States has not received marketing approval from the FDA and is only permitted under "enforcement discretion" for use in treating CDI that does not respond to standard therapy, there is a plethora of highly varied non-CDI indications for which FMT has been considered, and for which research is currently underway (Grigoryan et al., 2020). These include obesity, diabetes mellitus, Crohn's disease, ARO decolonization, allergies, inflammatory bowel disease, irritable bowel syndrome, ulcerative colitis, autism spectrum disorders, fluoroquinolone toxicity, mast cell activation syndrome, dysautonomia, chronic pain, chronic pouchitis, and combinations of complex medical disorders (Borody et al., 2013). Many studies are in progress, but at this point the available preclinical and clinical data is still insufficient to justify the use of FMT in non-CDI contexts. The gaps in understanding vary depending on the indication in question. For instance, with regards to multifactorial diseases such as diabetes and cardiovascular disease, in which disease pathogenesis often occurs over decades rather than the relatively limited timeframes of currently available FMT studies, additional research is necessary before the potential of FMT to contribute to treatment can be effectively assessed.

Finally, it is necessary to acknowledge that the aforementioned questions pertaining to mechanism of action, donor and recipient selection, and the standardization of screening and transplantation protocols may be indication-specific, further adding to the myriad of unanswered questions that remain regarding FMT.

Conclusion

In summary, despite predominantly positive perceptions by the general public towards fecal microbiota transplantation, there remain numerous unanswered questions surrounding this treatment, both with regards to its application in treating CDI and its potential applications for other indications. The recent rise in DIY adaptations of FMT and unregulated use outside of clinical supervision highlights the need for answers as the regulatory future of FMT remains unclear at this point in time.

How does the environment affect fecal transplants?

Introduction

Fecal microbiota transplants (FMT) and their interactions in the intestinal environment are what allows their transformative effects to occur. Healthy, functional gut microbiota can be introduced to a patient having decreased microbial diversity to alleviate symptoms or even cure a variety of disorders in the body. However, there are many other factors in the gut environment, the span of the body as a whole, and external social factors that can support or reduce the success of FMTs occurring in an individual.

Presence of Underlying Intestinal Conditions

As discussed, FMTs are used primarily in the treatment of Clostridium difficile infections (CDI). The degree of infection that the patient is inflicted with can directly affect the success of the FMT. Metronidazole and vancomycin are common antibiotics prescribed to treat CDI, having anticipated recovery times of under seven days (Castro et al., 2019). CDI may come in partially treated form, where a patient inflicted with CDI receives an appropriate treatment of antibiotics that leads to some infection alleviation, but does not experience a total loss of symptoms and curing of the infection. CDI can also be known as recurrent or refractory, where CDI that has been treated makes a recurrence in the patient in the long-term, often happening multiple times.

A 2021 study by Yoon et al. found that patients given FMT having partially treated CDI fared better and experienced a successful recovery at a higher rate than those with refractory CDI. Partially treated CDI had an exceptional recovery success rate of 100%, while refractory CDI was only 71.4% successful in comparison. Most notably, the lower alpha-diversity in the gut environment was restored in all successful cases "to a level similar to that in donor stools" (Yoon et al., 2021). This shows that the nature of the CDI influences the success of FMT procedures. Patients suffering from CDI may want to consider these findings from an alternative treatment of antibiotics first if they are unsure about FMT treatment at the moment or do not have access to it.

The Vagus Nerve

The microbiota-gut-brain (MGB) axis describes the interrelated link between the nervous system and the digestive system. Intestinal organisms and gut microbiota have been proven to regulate the brain and behaviour through the MGB axis in both direct and indirect pathways (Kim & Shin, 2018).

A study by Li et al. (2017) explored the influence of the vagus nerve on intestinal microbiota. In the study, the primary target was the treatment of a neurological infection, sepsis-associated encephalopathy (SAE), through application of FMT and vagus nerve procedures. The vagus nerve is of interest as it is the longest nerve in the body, and is one of two major pathways that transmit neuroimmune signals between the brainstem and abdominal organs like the stomach and intestines (Chaudry & Duggal, 2014). A vagotomy is a surgical procedure where selected branches of the vagus nerve are severed. There are three main types of vagotomy procedures - truncal, selective, and highly selective vagotomy (Seeras et al., 2021). In the study by Li et al. (2017), the vagotomy was performed by dissecting the left vagus nerves from the carotid artery. Rats were divided into different groups - a control group, a group receiving FMT, and a group receiving FMT along with the described vagotomy procedure. They found that in the FMT only group, FMT effectively reduced SAE symptoms and restored the gut environment of the rats closer to that of the donor compared to the control group. However, rats that received a vagotomy procedure in addition to FMT actually saw the positive effects of FMT reversed.

The results of this study show that the MGB axis must be considered when considering the use of FMT in patients. Negative effects of vagotomy interaction with FMT were not experienced in the study, only the reversal of the changes resulting from FMT. Since the vagus nerve has particular importance, patients may want to consider incorporating practices that benefit this nerve in order to support their physiological environment to best accept their FMT. Breit et al. (2018)

explained that vagus nerve activity can be stimulated through ordinary actions centred around relaxation, modulating the MGB axis on a whole. Activities such as deep-breathing exercises, immersion of the face in cold water, laughter, a foot massage, loud gargling with water to activate the vocal cords, and many more are some suggestions from Allied Health (2020) to stimulate the vagus nerve.

There are even medical procedures involving the surgical insertion of a vagus nerve stimulation device in the upper chest. This device is programmed in its electrical impulse duration and frequency according to the patient's specific situation (Howland, 2014). It is a common procedure for those with inflammatory bowel disease (IBD), but can also help with other neurological disorders such as chronic depression, headaches, Alzheimer's, and bipolar disorder (Mayo Clinic, 2020). These methods of vagus nerve stimulation may complement the application of FMT in the bodily environment to allow for relief of IBD and CDI symptoms.

FMT Contamination with Non-Bacterial Organisms

FMT is revolutionary in its ability to provide patients with healthy gut bacteria from donor fecal matter. However, the composition of FMT donation is not only microbes. In fact, bacteria make up a very small fraction of the feces transferred. "Faeces is a complex material of various biological and chemical entities that may be causing or assisting the effects of these treatments" (Bojanova & Bordenstein, 2016). Colonocytes, archaea, viruses, and fungi are some of the other living organisms that are introduced to a recipient's intestines through FMT. These non-bacterial organisms can have a significant effect on the biology of the FMT recipient.

Fecal matter contains the highest density of microbes in the human body. Bacteria makes up about 6.3% to 13.5% of fecal matter. The living, viable bacteria that is actually active is much lower, at about 50% of the mentioned composition percentages (Bojanova & Bordenstein, 2016).

Many viral bacteriophages are a smaller, but impactful, portion of FMT. There can be up to 109 viruses in a single gram of fecal matter (Bojanova & Bordenstein, 2016). These viral particles can act in a beneficial manner during FMT. Phages can target certain harmful bacteria in the intestines, preventing infections such as CDI, while avoiding useful ones. They can even reduce C. difficile cell density and toxin production (Meader et al., 2013), effectively combating the disease and offering an alternative to the aforementioned use of antibiotics.

Colonocytes are epithelial cells of the large intestine (Fonti et al., 1994). By covering the internal intestinal surface, they provide protection to nearby blood and tissues by acting as a permeability barrier that restricts gut microbiota movement. They also make up a portion of fecal material due to their shedding from the intestinal wall following cell death. In cases of colon damage caused by death of too many colonocytes, inflammation of intestinal tissue and development of bowel disorders can occur. Stem cell therapy was performed on a mouse colon model by Yui et al. (2012). The mouse had many regions throughout the intestinal tract that were deficient in colonocytes. Following the transplant of viable colonocytes stem cells into the mouse, colonocytes easily formed epithelial layers in the regions lacking them prior. This development of colonocyte layers restored the protection of the intestines and reduced inflammatory disorders. This poses options for human FMT containing viable colon stem cells. Treatments like this aid recipients lacking the protective colonocyte layers. The gut dysbiosis present in diseases resulting from disrupted colonocyte metabolism, such as IBS or IBD (Litvak et al., 2018), could be ameliorated in this way through the coordination of an FMT environment containing a sufficient concentration of viable colonocytes.

Diet and Exercise
Gut microbiota can also be influenced through a patient's diet and exercise. Many health food experts in popular culture describe the importance of fuelling the body with foods that improve gut microbiota to combat intestinal inflammation. Studies have been done in observing the inclusion or exclusion of foods in one's diet to support the mechanisms of FMT in combating illnesses such as CDI, obesity, constipation, and more.

A diet including soluble dietary fibre can greatly complement FMT in the treatment of slow transit constipation (STC). Fibre digestion on its own regulates intestinal microecology through its interaction with gut microbes and support of production of metabolic products. An increase of just 16 grams of soluble dietary fibre per day for 4 weeks following FMT was shown to improve assorted STC symptoms such as quantity of bowel movements per week and colonic transit time (Ge et al., 2016). Ge et al. compared their findings to a past study by Tian et al. (2016) where FMT alone was used to treat STC. The combined treatment of dietary fibre and FMT showed to improve symptoms of constipation significantly versus the solitary FMT.

Scientists have also considered the modification of the FMT donor's environment. While FMT can help alleviate some of the patient's symptoms on its own, these studies show that the accompaniment of a modified diet provides a better environment through which the FMT can act upon. A 2020 study by Zoll et al. looked at the effects of the gut microbiota on obesity. They worked with mice subjects as donors splitting the mice into different groups based on their environmental actions. One group was fed a calorically-dense, high-fat and high-sugar diet, while the other group was fed a normal diet. These groups were further divided by their exercise and activity. The mice in each group either maintained a sedentary activity level while on their assigned diet or an active level by running on a treadmill each day during the course of the 6-week study. Following the adoption of the lifestyles, FMT was performed from the donor mice to recipient mice. Zoll et al. found that while the FMT diet did not modify the recipient's body composition, it greatly affected their ability to metabolize glucose and other macromolecules. Insulin levels of recipients of the high-fat, high-sugar diet were disrupted, leading to a decrease in the ability to metabolize glucose.

Based on their existing lifestyle, patients or medical professionals may want to consider different donors. While the gut microbiota of one individual is healthy and functional in their own body, it may not translate well as a donor through FMT if the recipient's diet is significantly different. Patients can also consider modifying their diet to more easily digestible foods, or to resemble the makeup of the donor's diet. Certain diets such as the specific carbohydrate diet (SCD) have proven to alleviate symptoms of Crohn's disease, a type of IBD (Arjomand, 2020). This includes a regimen of eliminating sugar, grains, and starch from daily consumption in the patient. Situations where glucose metabolism is impaired due to high-sugar donor diets, such as those seen in the study by Zoll et al., may be avoided with the adoption of SCD. Recipients can still experience the benefits of good microbes received through FMT while evading dysfunction of glucose metabolism by simply restricting its consumption.

Exercise has been shown to cause changes in the gut microbiome.

In the previously mentioned study by Zoll et al. (2020), exercise-induced effects experienced by the exercise-trained donors did not translate into the FMT recipients. However, "exercise independently alters the composition and functional capacity of the gut microbiota" (Mailing et al., 2019). Exercise in a patient has the ability to change the composition and function of the gut microbiome, independent of the individual's diet. Mailing et al. (2019) describe that in women, performing at least 3 hours of exercise per week increased their levels of various gut bacteria, including Akkermansia muciniphila, compared to the control inactive women. A. muciniphila is known to be beneficial in reducing markers of obesity such as adiposity, waist-to-hip ratio, and body fat distribution (Dao et al., 2015). Patients receiving FMT for obesity treatment should consider the adoption of physical activity into their lifestyles. From these combined efforts, they can experience improved digestive effects due to the donor microbiome as well as the introduction of a higher level of metabolic bacteria from their exercise regime.

Social Environment and Patient Decisions

While much of the physiological environment's effects in relation to FMT have been touched upon in this chapter, there is an inherent social environmental aspect that also affects FMTs. As will be discussed later in the book, there is controversy surrounding FMTs and they are displayed a certain way in popular culture as well as the medical field. Due to this social element, eligible patients may perceive a stigma and aversion in regards to FMTs. Medical professionals themselves may also have a personal perception that influences their choice to recommend FMT to their patients.

A study by Park et al. (2017) examined social or personal factors that influence patients to accept FMT procedures. While knowledge of FMT was low among the surveyed patients at only 12%, there was still a high percentage of patients "willing to undergo the procedure if medically indicated," at a response rate of 77% (Park et al., 2017). There were many lifestyle factors and characteristics that ultimately caused patients to undergo FMT. Marital status, having children, and education level were some significant aspects studied. Those who were married were more likely to accept FMT than unmarried individuals (Park et al., 2017). Additionally, patients with children were also more inclined to FMT than those without. This is believed to result from the nature of familial relationships. Park et al. suggest that it is because these patients are driven to provide for their family. An unattractive procedure such as FMT may be more appealing to those with families as improving their health means that they can be around longer to care for their family, and are more physically capable of doing so. Other life history can impact patient decisions as well. Patients that have previously been hospitalized are more willing to undergo FMT (Kahn et al., 2013), likely stemming from their trust and comfortability with the medical field.

Patients may also be concerned and refuse FMT if related risks are not outlined. Main concerns among candidates recommended for FMT include factors such as "adequate screening for infections, cleanliness, and potential to worsen illness" (Kahn et al., 2013). This relates back to the social stigma and perceived unappealing nature of using another individual's feces as a medical treatment.

Patient willingness to undergo novel procedures such as FMT is important for the future of medicine. Increased use of the practice will provide more experimental data on the use of FMT in various conditions and their efficacy in associated symptoms. It is important that patients are knowledgeable and comfortable with these procedures before accepting to undergo them, outlining the emphasis on communication in healthcare from medical professionals.

Familial Relation and Social Interaction

A basic understanding of genetics tells us that the genotypes of closely related individuals are more similar to each other than those who are unrelated. Biological research has shown that the genetics of the host are a major influence in the development of the gut microbiome (Goodrich et al., 2014). Gough et al. (2011) found through a systematic review that FMT performed where there was a close relationship between the stool donor and recipient showed greater relief of symptoms in CDI. The CDI resolution rate was 93% for related donor FMT, such as family members or spouses, while slightly lower at 84% resolution for unrelated donors. Also notable was the discovery that FMT donated from a spouse or partner had a higher resolution rate of CDI than donation from a family member. These results may indicate that both genes and lifestyle can impact the acceptance of FMT and the integration of donor microbiome into the recipient. Related family members will likely have similar phenotypic expression of microbes in their intestinal environment due to their shared genetic similarity. Dill-McFarland et al. (2019) have shown that close social relationships can also impact the composition of the gut microbial community. This was proven not only in human interactions, but also in animals such as wild baboons. Living together with a roommate may have an impact, but the quality and intimacy of the relationship is more dominant in determining the similarity of microbial diversity between individuals.

These findings do pose a restriction in many cases. Not all patients have closely related individuals in their lives available to donate stool for FMT. While Gough et al. (2011) suggest that "relatives of patients could be given priority as potential stool donors," this may not be possible in every case. Patient relationship status may limit their connection with a spouse or partner as a potential stool donor, they may not have close social relationships, or their citizenship and residence in another country may cause a physical distance barrier between themselves and relatives.

Conclusion
Many environmental factors can influence FMTs. While some factors are beyond control of the individual, it is clear that candidates for an FMT should closely consider the actions they take in their lives that may influence their gut microbiota, such as dietary intakes or preexisting conditions. Proper knowledge translation from medical professionals to their patients should also occur to ensure patients are given an unbiased and informative discussion regarding FMT to clear up any inconsistencies in their knowledge and allow them to make an informed decision on whether to undergo an FMT procedure. FMT is an innovative pathway in the treatment of severe illnesses as mentioned earlier in the book, such as CDI, metabolic syndromes, and IBD (van Nood et al., 2014). Before undergoing the treatment, patients may want to consider some of these factors in their everyday lives that may affect their FMT procedure for better, or for worse.

What controversy is there surrounding fecal transplants?

Introduction

Fecal microbiota transplantation (FMT) is a new technology that has gained interest in the past couple decades. Drugs account for 85% of total costs in a German outpatient setting so the need for an efficient and cost-effective alternative is one of the reasons health professionals may turn to FMT (Zeitz et al., 2017). There is increasing interest in treating the infection Clostridium difficile (C. difficile) causes with FMT, by recolonizing the gut (Alagna et al., 2019). This infection is contracted in hospitals and other health care facilities like long-term care homes. The incidence of C. difficile infections has steadily increased in the past two decades (Rubin et al., 2018). Antibiotics are losing their effectiveness and patients have experienced multiple recurrences of the infection (Sachs & Edelstein, 2015). The cure rate is 90% for FMT, whereas antibiotics are only 30-40% efficient, so FMT is slowly becoming more common (Sachs & Edelstein, 2015).

Benefits

FMT is not a preventative method; it is only used after the onset of the disease (Rubin et al., 2018). FMT is used to treat the Clostridium difficile infection (CDI) effectively, yet is still a scarcely used option (Alagna et al., 2019). Evidence on the benefits of FMT is controversial and inconclusive (Zeitz et al., 2017). Alagna et al. (2019) did observe that all patients recovered from CDI after FMT. However, some patients had already shown signs of recovery prior to FMT initiation so there is a possibility that recuperation may have happened without FMT.

While it has shown success, the lack of data on long-term safety of FMT has no consistency (El-Matary & Wael, 2013). The majority of patients who have undergone FMT have not had long-term follow-ups and case studies that have studied these effects usually only have a few participants (El-Matary & Wael, 2013; Ma et al., 2017). The long-term safety of fecal transplants in children has not been adequately studied even if it has shown promising results in adults. In a study involving 10 patients below the age of 18, 90% recovered from CDI (Nicholson et al., 2014). However, children are still developing and their microbiomes would be different from the composition of an adult donor's. Altering children's microbiota may have unforeseen effects as they mature.

Effectiveness depends on the parasite/pathogen being eradicated. For Klebsiella pneumoniae (K. pneumoniae) FMT only caused partial eradication (60%) whereas for Escherichia coli, it had 100% eradication (Bilinski et al., 2017). K. pneumoniae eradication needs Barnesiella, Bacteroides, and Butyricimona species in the fecal matter, so the donor's microbiome in the gut matters. In another study, patients receiving immunosuppressant therapy achieved better outcomes than just FMT alone, indicating that FMT may not work as well on its own (Zhang et al., 2020). Likewise, many studies have used patients who have already been given antibiotic treatment prior to FMT, agreeing with the previous statement that success may not be due to FMT alone (Van Nood et al., 2014).

Data is most documented for CDI, but for other diseases, like ulcerative colitis and Crohn's disease, there is limited research on FMT's benefits (Alagna et al., 2019). For ulcerative colitis, many treatments may be needed for FMT (Nishida et al., 2016). For Crohn's disease, research has only been done on very small focus groups (Padma, 2019). Like ulcerative colitis, multiple doses of FMT may be required as the disease can spontaneously relapse and a single dose has a modest effect.

A recent endeavour, FMT is used to treat inflammatory bowel disease (IBD) (Shanahan & Quigley, 2014). Unfortunately, it only works when the microbial colonization of the gut is just starting and the immune system is still in the early development stage. This is a very narrow window to diagnose and treat the disease. The immune system has a strong influence on the development of the internal microbiome. There are concerns with manipulating microbiota in patients with innate immune defects for this reason. If the immune system continually influences the development of the microbiome, what the FMT based therapy accomplishes may be undone in the long-term. In addition, the effect of IBD is not always predictable . Twelve percent of patients treated required hospitalization 12 weeks after FMT, whereas the rest did not experience any serious adverse effects (Fischer et al., 2016).

Physicians remain divided; some are outright skeptical of how beneficial it is, mainly due to the ease of transfer of pathogens in bodily fluids (Glauser, 2011). Dr. Andrew Webb, Vice President of Medicine at Fraser Health, says the evidence is not strong enough to support the widespread use of FMT, while other physicians have accepted the science for treating CDI (Glauser, 2011). Whether it will continue to be used for other diseases is still in question.

Donor screening

The optimal donor choice for FMT involves a variety of factors and the ideal criteria to screen for healthy donors are unclear (Zeitz et al., 2017). Sometimes, patients are uncomfortable with identifying suitable donors (Alagna et al., 2019). In emergency situations, patients cannot choose their own donors so frozen stool banks are needed. One aspect to consider is that relatives may not be suitable donors if the goal of FMT is to modulate the recipient's microbial composition, since immediate relatives are likely to share genes and/or a similar diet (Nishida et al., 2017).

The treatment of chronic disease with FMT is dependent on the diversity and composition of the donor's microbiota (Wilson et al., 2019). The donor's diet and genetics play a factor since no two gut microbiomes are alike. This results in a range of clinical responses for the patient. Gastroenterologists have hypothesized that creating a donor-patient matching system would create a better response since there is no "one donor fits all" approach (Wilson et al., 2019). In a study by Nishida et al. (2017), the Japanese microbiome gut was found to be different from the western one due to diet, so efficacy and safety of FMT will vary with the location and nutrition of the individual. In cases of FMT failing, one of the reasons for the lack of success is due to the reduced immunomodulatory properties of donor feces (Alagna et al., 2019). Donor feces that contain a certain amount of butyrate-producing anaerobic bacteria like Faecalibacterium prausnitzii, Roseburia hominis, and Coprococcus eutactus are found to have the most success. The mechanism behind this is still being determined.

For these reasons, it is very hard to find eligible donors. There is no well-defined age restriction; China has used children aged 6-24 as the ideal target for donors, whereas other studies have a strict policy of being adults (Zhang et al., 2019; Paramsothy et al., 2015). Healthy donors had to be between 18-65 years of age in a study done by Paramsothy et al. (2015). Of a pool of 116 potential donors, only 10% of those screened ended up donating a sample . Many were unwilling or unable to meet the commitment requirements, with factors involving BMI, medication use, or drug addiction. Those who had colorectal cancer or a family history of it could not be considered eligible either, since the transfer of the gut microbiome could also transfer the disposition of the cancer's development to the recipient (Paramsothy et al., 2015; Stallmach et al., 2020).

Another area of debate is the criteria for screening. Having gastrointestinal parasites is also grounds for exclusion; the most common parasite discovered is Dientamoeba fragilis (Paramsothy et al., 2015). Screening for pathogens is done with careful scrutiny. Acute infections can arise even with pathogen concentrations below the technical threshold of detection and in the absence of a serologic response (Stallmach et al., 2020). Donor screening heavily relies on the honour system, where donors would have to disclose any conflicts of interest and adhere to rules of eligibility, like drug usage. Individuals who may be exposed to a pathogen after initial screening would have to disclose the event. Therefore, researchers suggest a second screening 8-12 weeks after the first test. This could pose a problem in emergency cases where the patient's condition could rapidly deteriorate in 2-3 months (Zhang et al., 2019).

The number of institutions that offer FMT are limited, so patients and potential donors may have to travel considerable distances. Whether or not the journey is worthwhile is still being analyzed. Since few hospitals have the right equipment to do them, fairness and accessibility issues are brought up (Glauser, 2011).

Method of Preparation

There are many limitations to the method of preparation since there is no standardized procedure for doing so (Zeitz et al., 2017). How the microbiota is prepared influences the development of adverse effects, which usually happen one month after the transfer (Zeitz et al., 2017; Ding et al., 2019). Preparation for the colonic transendoscopic enteral tube route results in lower adverse effects than other routes for treating CDI (Ding et al., 2019).

The development of an automatic preparation technique, using GenFMTer, is also a way to lower adverse effects, more so than manual preparation (which involves filtration plus centrifugation). Specifically for Crohn's disease, manual and automatic purification of fecal microbiota did not correlate with the efficacy of FMT (Wang et al., 2018). Only 14% of patients with Crohn's disease had mild adverse effects that did not last beyond one month. Therefore, preparation may be important for CDI but not for other diseases.

There is still no agreement on the best protocol for FMT preparation or delivery (Van Nood et al., 2014). It has been suggested that preparation be done in an anaerobic environment to preserve the obligate anaerobic bacteria, which is mentioned above as lending effectiveness to FMT. While this may be true, most labs lack the anaerobic chambers to do so.

Delivery process

There is no consensus by professionals on the ideal route of FMT (Zeitz et al., 2017). The timing and duration of administration required to cause and maintain a clinical response are also unknown. Different routes will have a varying risk of adverse effects. In a study on those suffering from CDI, critically ill patients who received FMT via a nasogastric tube, as opposed to enema or colonoscopy routes, experienced higher rates of serious adverse events such as fever and aspiration pneumonia (Alagna et al., 2019). This agrees with a study done on the colonic transendoscopic enteral tube route that showed lower adverse effects (Zhang et al., 2020). The newest routes are the fecal capsule, lyophilized frozen stool, and fecal filtrate transfer (FFT); their efficacies are not entirely known as more clinical trials need to be done (Alagna et al., 2019). Sterile fecal filtrate has been shown to be effective in preventing recurrence of CDI (Rubin et al., 2018). It is also less cumbersome and safer to deliver (Stallmach et al., 2020). Capsules and lyophilized stools do not have as standardized a process as the others do.

For diseases other than CDI, fecal enemas are tolerated by children with ulcerative colitis (UC), who have only mild adverse effects like cramping, bloating, and diarrhea (Kunde et al., 2013). Effects are self-limiting, so children can recover from effects without medical intervention. Thus, the ideal delivery route depends on the disease and the person being treated.

Laws and Policies

On the research side of the FMT, ownership of the gut microbiome is debated (Hawkins and O'Doherty, 2011). Privacy when disclosing test results is a significant matter. Microbial biome makeup is akin to identification of the person via DNA or fingerprints. Samples are of a highly personal matter, affecting human dignity, as well as cultural sensitivities; consideration of whether the sample can be linked to a certain individual must be done. Other scientists have argued that microbial genomes are not part of the human genome and that microbial samples are considered waste, whereas donations of blood are not (Hawkins and O'Doherty, 2011).

FMT started out as a fringe medical practice in the late 1950s and has only recently gained more acceptance with the growing resistance of microbes to antibiotics. Originally, the Food and Drug Administration (FDA) barred FMT's to everyone except those in clinical trials. They eventually allowed FMT for all patients, so long as the treating physician had the patient's consent. Guidance for physicians doing or recommending FMT is another facet.

There is commercial value in donor feces. Its lucrativeness has caused the FDA to decide to regulate fecal matter as a drug. Laws and regulations are necessary given that there is a risk of stool-borne transmissible illness and determining accountability is needed, should this matter arise (Sachs & Edelstein, 2015). Some physicians and patients have tried a "do it yourself" approach to donor acquisition and preparation; donors not properly screened can exacerbate circumstances. The problem is that FMT is not the same thing as small molecule drugs. It does not have an active ingredient; effectiveness and safety are related to the life history of donors, and composition and quality are highly variable. Currently, there is no patient registry to track the adverse effects of FMT (FDA, 2019).

Treating FMT as a drug means it can fall prey to the capitalist market (Sachs & Edelstein, 2015). Drug prices are dependent on the market and companies have a monopoly on supply and distribution. Inflation is rampant for drugs like colchicine, which used to be widely available and cheap until one company formed a monopoly and forced all other makers to exit the market. FMT donation banks could go out of business should they enter the market and prices will rise dramatically. Companies that try to commercialize FMT products cannot have the same approach as antibiotic regulation since FMT is by donation and they are not made in the lab like antibiotics can be (FDA, 2019). The FDA has had to issue an alert after an FMT resulted in death during a clinical trial. The FDA has amended their recommendations once it became known that multidrug resistant organisms (MDRO) could spread their colonization through FMT (Soucheray, 2019). MDROs will decrease the effectiveness of antibiotics, which is the very thing FMT is trying to combat. Thus, donors must be screened for possible MDROs, not just intestinal parasites.

What do patients think?

The lack of patients' knowledge of FMT as a therapeutic option is another hurdle for the widespread use of this remedy (Zeitz et al., 2017). In a survey, patients' views were mixed; they preferred FMT equally to traditional medication (about one-third of responders each). There is fear surrounding the procedure, stemming from the risk of garnering infectious disease. Disgust was also prominent, although some people do not know that FMT can be done orally through a capsule. Half the people who responded to the survey indicated that a discussion with a specialist would likely change their minds. Patients favoured transplantation via colonoscopy and 38.8% preferred a family member as a donor. The best delivery method depends on the patient, where nausea and vomiting are a factor in determining the ideal delivery route (Zhang et al., 2019).

Acquiring informed consent from vulnerable patients is the most challenging step regarding FMT (Ma et al. 2017). Even if attitudes are generally open-minded, there are still low levels of patient knowledge on the process and a lack of awareness of the published literature available (Ma et al. 2017). In countries outside of North America and Europe, gastroenterologists and other physicians have limited experience with FMT since most of the research has been done in Western countries.

Summary

Most of the controversies surrounding FMT stem from the unknown. The method of preparation, ideal donor, and the ideal route of transfer differs between those with CDI and the other diseases mentioned, so more research is desperately needed (Kunde et al., 2013). In fact, most of these studies have very small populations looked at thus, research is very limited, making most of the suggested therapeutic benefits skeptical (Shanahan & Quigley, 2014). The debate on the safety, necessity, and details of FMT will continue as more clinicians and researchers turn to it as a viable alternative to antibiotics.

How are fecal transplants portrayed in popular culture?

Introduction

Fecal transplants, more formally called fecal microbial transplants/ fecal microbiota transplants (FMT), or stool transplants, emerged in the year 2003 as a health intervention used to treat issues pertaining to the gastrointestinal tract (McLeod et al., 2019). More specifically, fecal microbiota transplants have been defined as procedures in which fecal matter is retrieved from a donor who has been carefully tested (McLeod et al., 2019). The collected fecal matter is then combined in a saline or another solution, strained out, and inserted inside the body of a patient via various procedures, including colonoscopy, endoscopy, sigmoidoscopy, or edema. This procedure is performed in patients who have had their gut microbiota destroyed or adversely impacted as a consequence of a multitude of reasons, including, but not limited to, infection, antibiotic treatments, and chronic intestinal disorder (McLeod et al., 2019).

This chapter will deconstruct how fecal microbiota transplantation is perceived in popular culture. Specifically, it will briefly highlight how fecal transplants have been used and perceived in the past, as well as how patients reacted to the concept of fecal microbiota transplantation when implemented in relation to themselves or their loved ones. This chapter will also briefly discuss how fecal microbiota transplants have been, and continue to be portrayed in the print media, as well as overarching themes that the media tends to gravitate towards when introducing the concept of fecal microbiota transplants to the greater community at large. Lastly, the stigmas associated with fecal matter throughout history, as well as those that are present today are also highlighted, with an emphasis on how healthcare providers can be affected by these stigmas, and thereby serve as a barrier to patients who may benefit from a fecal microbiota transplant.

A brief history on the origins and preliminary perceptions of fecal microbiota transplantation

In popular culture, it has been difficult for fecal microbiota transplants to gain public acceptance as a legitimate, mainstream health intervention to treat disease. In fact, although the procedure was first brought to light globally in 2003, via a Scandinavian report that proposed fecal microbiota transplantation as an alternative treatment for Clostridium difficile (C. difficile) infection (St. Joseph's Health Care London, n.d.), treating diseases of the intestines with fecal matter has been used in historical medical practices across the world (Klasco, 2019). In fact, fecal microbiota transplantation was first implemented in fourth century China to treat illnesses concerning the gastrointestinal system; the Chinese would use "yellow soup" to treat food poisoning, diarrhea, and later, even fever and pain (Zhang et al., 2015).

Perhaps one of the most fundamental moments in the history of fecal microbiota transplants did not occur such a long time ago, but rather in 1958, when doctors used the procedure sporadically to treat four patients suffering from life-threatening C. difficile infections (Klasco, 2019). However, despite all four patients surviving, and a 94% success rate among 16 more patients who underwent the same procedure during the next twenty years, fecal microbiota transplants were unable to gain acceptance in mainstream medicine (de Groot et al., 2016).

The "yuck factor" regarding fecal microbiota transplants

Kahn et al., (2013) describe this apparent unwillingness for patients to consider fecal microbiota transplantation as the "yuck factor". In their study, adult patients and parents of children with ulcerative colitis (UC) and indeterminate colitis were asked about their perceptions and interest in undergoing (or having their children undergo) fecal microbiota transplantation as a treatment for their condition. In their study, all but one subject was interested in fecal microbiota transplantation as a treatment for themselves or their child, with parents of children expressing less concerns and reservations than adult subjects. One patient expressed that "the initial thought is the 'yuck factor' but ...[they were] ... at a stage in [their] colitis where [they were] ... on the last straw", whereas another stated that fecal microbiota transplantation "sounds disgusting but depending on what stage [a person] is in [their] illness, a patient might be potentially willing to overlook the disgusting nature of [the procedure] just to get relief" (Kahn et al., 2013).

Further, although the individuals in Kahn et al.'s (2013) study were generally willing to consider fecal microbiota transplantation as a treatment option, many concerns were raised about the risks and general logistics of the procedure. For instance, all subjects agreed that they would prefer enema or colonoscopy as a delivery method for the transplant, rather than via nasogastric tube. Further, subjects generally agreed that their willingness and desire to participate in fecal microbiota transplantation (for themselves or for their children) would be majorly influenced by how much support their healthcare providers would provide them regarding the treatment.

Interestingly, branding for the treatment name was a crucial factor in a patient's desirability towards the procedure (Kahn et al., 2013). Some individuals felt that using the word "fecal" in the treatment name made it less desirable overall; in essence, this encapsulates the "yuck factor" Khan et al. (2013) described.

Perhaps the most interesting facet of the study however was highlighted when Kahn et al. (2013) asked subjects about their social concerns pertaining to fecal microbiota transplants. Most of the patients did indeed exhibit disgust and aversion to the concept of fecal transplantation; however, this distaste did not affect how interested subjects were in fecal microbiota transplants because generally, all subjects were in agreeance with the notion that colitis itself is a difficult, unpleasant experience (Kahn et al., 2013). In essence, subjects felt that living with colitis (or in the case of parents of patients, seeing their children live with colitis) changed their perceptions towards fecal microbiota transplants, and as a result, all subjects agreed that they did not feel that fecal microbiota transplant was a "dirty" or unhygienic procedure (Kahn et al., 2013). Further, all patients agreed that fecal microbiota transplantation did not clash with their religion or cultural background (Kahn et al., 2013).

However, social stigma was still an apparent issue, despite subjects' willingness to undergo the procedure or have their children undergo it (Kahn et al., 2013). In fact, adult patients who already preferred privacy in regards to their condition said that it was unlikely for them to share their fecal transplant journey with their friends and family, with many mentioning that they expected their friends and family to express jokes about their treatment (Kahn et al., 2013).

Fecal microbiota transplants: portrayal in the print media

English language newspapers discussed fecal microbiota transplants for the first time ever in the year 2003; this was in the Australian Financial Review, in an article that depicted the procedure as a last resort treatment for C. difficile. Although the article did mention that fecal microbiota transplantation can be a "plain distasteful" procedure for many individuals, it also highlighted that various other medical procedures, such as blood transfusions, were subjected to the same distaste from the public.

McLeod et al. (2018) performed a study in which they searched for keywords pertaining to fecal microbiota transplants, such as "fecal microbial", "microbiota transplant", and "stool transplant" on a UK news database. Upon evaluating the results, it was determined that there were three major themes in relation to how fecal microbiota transplants are portrayed in the print media. The first theme was the challenging of how feces are represented in society, in attempts to make readers of print media familiar with associating feces as medical terminology, rather than something that evokes disgust or aversion. Secondly, the print media often employs the use of metaphors to relate the more emergent, recent concepts of microbiota and the microbiome to older concepts, such as probiotic bacteria, which have a general positive connotation. The third and final theme observed was the attempt to associate the newer concepts of transplantation for medical purposes to other, older medical technologies that members of society are more familiar and comfortable with.

These three themes were depicted in future representations of fecal microbiota transplants in the print media. Although there was nearly a decade-long gap in reporting the procedure in English language newspapers since 2003, headlines from 2012 and 2013 encapsulated fecal microbiota transplants in a variety of ways (McLeod et al., 2018). For example, the procedure was first linked to the microbiome in 2013 (McLeod et al., 2018). This was on-trend for the time, as the early 2010's saw a massive rise in interest concerning the microbiome and fecal matter as "an industry" (McLeod et al., 2018). All of this did indeed lead to a gradual acceptance of fecal microbiota transplantation as a mainstream procedure. In fact, 2012 was the first time the concept of "do it yourself" fecal microbiota transplants were introduced to the public via the print media; this was a watershed moment because it conceptualized fecal transplants as something quite literally close to home, and contributed to print media's attempts to deconstruct the "yuck factor" associated with fecal matter in medicine (McLeod et al., 2018)

Where does the stigma associated with fecal matter, and thus, fecal microbiota transplantation, arise from?

The stigma associated with feces has been entrenched in the core of western society for decades. Mary Douglas (1966) stated that feces are merely just "dirt out of place"; while they are acceptable inside the human body, they become impure, unhygienic, and thereby dirty when discussed in context outside of the body (McLeod et al., 2018). Similarly, Sjaak Van der Geest (2007) reinforced this notion, suggesting that fecal matter is intimate, private, and should not be discussed or written about, because such actions "disturb the order of proper behaviour" .

Although it has been decades since the times in which Mary Douglas wrote Purity and Danger, her book on pollution and taboo, the mindset she proposes is still held widely by many individuals today, who see fecal matter as something disgusting, rather than potentially scientific. As a result, the idea of feces alone is distasteful to many, let alone the concept of transplanting a donor's fecal matter into another individual's body (McLeod et al., 2018).

As a consequence of this stigma, it seems that the major limitation for fecal microbiota transplantation therapy ultimately ends up being the acceptance of the procedure among patients and even physicians who are responsible for patients' suffering from gut conditions (Orduna et al., 2015). Orduna et al. (2015) used social media platforms, such as FaceBook and Twitter, in order to survey Mexican graduate students about fecal microbiota transplants. They found that most individuals surveyed were not familiar with the concept of fecal transplantation, and further, were hesitant to answer whether they would receive fecal transplants, or donate stool for someone else to use.

Orduna et al. (2015) therefore highlighted two major concerns surrounding fecal microbiota transplantation. The first concern was that due to a lack of education and information, individuals express hesitation and uncertainty in regards to both stool donation for another individual, and potentially undergoing a fecal microbiota transplant themselves. However, a positive outcome that was realized through this study was that 86% of the population that was surveyed confirmed that they were interested in learning more information pertaining to fecal microbiota transplants. The second concern that was raised through the study emphasized the idea that there must be new and improved avenues of communication, in order to facilitate the transfer of medically relevant information to the greater community at large (Orduna et al., 2015). For example, although it may initially be assumed that social media is a good way to share information about fecal microbiota transplants, a major drawback of using social media is that the majority of individuals who use it tend to be younger (Orduna et al., 2015). Thus, broader methods of communication, or many more narrow methods, must be implemented to raise awareness and education, thereby deconstructing negative associations with fecal microbiota transplantation.

Perceptions that patients hold in regards to the implementation of fecal microbiota transplantation

Generally, fecal microbiota transplants are portrayed as a more "natural" or even an "organic" method by patients who need to undergo the procedure (Kahn et al., 2013). As a result, it is also perceived as safer than antibiotics, and other conventional therapies (Kahn et al., 2013). However, although the United States Food and Drug Administration (USFDA) regulates human fecal matter as a drug and biological product (Kahn et al., 2013), critics deem fecal microbiota transplantation procedures as far from ethical. The most pertinent concern in relation to the procedure usually entails the process by which donors are selected and screened, as well as how vulnerable the patient is, and what the long term safety concerns are for patients who have undergone fecal microbiota transplants (Ma et al., 2017).

Perceptions that healthcare providers hold in regards to the implementation of fecal microbiota transplantation

Many major criticisms of fecal microbiota transplants arise, unfortunately, from the physicians themselves. Physicians and gastroenterologists attitudes in regards to fecal microbiota transplants were found to be conservative or negative, due to lack of exposure towards the notion of using fecal matter as a health intervention (Brandt, 2012). Further, it was found that clinicians tend to only consider fecal microbiota transplantation as a potential treatment when it is confirmed to be justified by science and is deemed ethical (Moossavi et al., 2015). However, it is important to note that while these are the general opinions of clinicians in western countries across North America and Europe, there is a lack of survey data from non-western healthcare providers pertaining to their opinions on fecal microbiota transplants (Ma et al., 2017). Ma et al. (2017) suggest that the ethical dilemma between patients and healthcare providers highlights a need for there to be more investigation into the clinical side of the procedure - specifically, the perceptions of the clinicians themselves - in order to implement fecal microbiota transplants in practice most effectively for the benefit of the patient.

Conclusion

The perception of fecal microbiota transplant in popular culture has undoubtedly changed greatly over the past decades. However, before the procedure can be accepted as a mainstream health intervention by the general public, it is fundamental for both clinicians, as well as patients to be well educated and aware of how fecal microbiota transplants work. Further, there must be more transparency in regards to the donation process, how the actual transplantation occurs, and the health outcomes for patients who undergo the procedure. Nonetheless, with the microbiome emerging as a hot topic in biological science, and clinicians looking for novel, creative ways to treat bothersome ailments, perhaps fecal microbiota transplants are a step in a new direction in regards to modern medicine.

What future directions is fecal transplant research moving in?

Super Donors

Several studies suggest that FMT's success depends on the diversity of microorganisms in the stool donor, leading to the idea that there may be FMT super donors (some stool is more effective at producing improvements than others) (Wilson et al., 2019). Further evidence is provided when FMT recipients experience an increase in stomach microorganism diversity, typically shifting towards a composition similar to their stool donor (Wilson et al., 2019). Individuals who have a higher response to FMT typically show more microorganism diversity than those who do not (Wilson et al., 2019). This has led to the suggestion that the key to FMT success lies in the ability of a donor to transfer high levels of a particular species (of microorganisms to recipients) (Wilson et al., 2019). But the concept of a super donor may not be overly beneficial. The results of an FMT trial with combined stool were not better than trials with stool from single individuals (Wilson et al., 2019).

Another potential approach involves donor-recipient matching, where a patient is screened to identify the specific problems in their stomach microbiome (Wilson et al., 2019). Then the patient can be matched to a specific FMT donor that is known to be rich in the species the patient requires to restore pathways in the patient's stomach microbiome. In addition, immune screening may also be beneficial to help reduce the impact of donor-recipient incompatibilities that result from differences in innate immune responses (Wilson et al., 2019). This approach could maximize FMT treatment success while avoiding common problems such as inappropriate immune responses.

Ulcerative colitis

Ulcerative colitis is a chronic illness characterized by inflammation and ulcers in the colon (large intestine) and rectum (MayoClinic, 2021). Ulcerative colitis falls under the category of inflammatory bowel disease (IBD). Individuals with ulcerative colitis can also have a Clostridium difficile infection (CDI), which can complicate and worsen the course of ulcerative colitis (Reinink et al., 2017). Individuals with ulcerative colitis have similar symptoms to Clostridium difficile colitis, such as diarrhea and stomach cramps (Reinink et al., 2017). Since the symptoms of ulcerative colitis and Clostridium difficile colitis are similar, correctly diagnosing a patient presenting with diarrhea and cramps becomes difficult for clinicians (Reinink et al., 2017). An accurate clinical diagnosis of CDI is also difficult, as symptoms can be confused with an acute (short-term) flare of ulcerative colitis (Reinink et al., 2017).

Recently, polymerase chain reaction (PCR) assay for Clostridium difficile (a bacterium; hereafter referred to as C. diff) is used to test for infection (Reinink et al., 2017). PCR assays test for infection by examining a small sample of an individual's DNA to determine whether it contains genetic material from C. diff. Such PCR tests for C. diff have high sensitivity but this comes at the cost of detecting a large amount of colonized cases of the bacteria, in which case C. diff is only a bystander and the underlying problem of active ulcerative colitis may go unnoticed (Reinink et al., 2017). Thus, this highlights a necessity for having better tools to differentiate between CDI and an ulcerative colitis flare, especially when the PCR assay is positive (Reinink et al., 2017).

This difficulty in diagnosis presents a clinical dilemma while awaiting the results of a C. diff test. Should clinicians start a patient on an antibiotic course to address CDI in those with diarrhea, recent antibiotic exposure, and without other intestinal pathologies (Reinink et al., 2017)? If an individual is truly infected with C. diff, the answer is obvious they should start treatment right away (Reinink et al., 2017). But there is an added complexity to this question in ulcerative colitis. Since the symptoms of CDI have great overlap with those of an acute ulcerative colitis flare, choosing the appropriate treatment is crucial for patient health outcomes (Reinink et al., 2017). The treatment of choice for ulcerative colitis is immunosuppression, which involves various types of medications (Reinink et al., 2017). But treatments that increase immunosuppression are contraindicated (having a condition such as CDI serves as a reason to not undergo a certain treatment because it would cause harm) for addressing untreated CDI (Reinink et al., 2017).

On the other hand, waiting to begin immunosuppression treatment in a severe ulcerative colitis flare can increase an individual's chances of dying from the disease (Reinink et al., 2017). The time it takes to rule out CDI as a cause of worsening symptoms could explain the worsened long-term outcomes of patients who get a CDI (Reinink et al., 2017). This is why the PCR assay and diagnostic methods need further research. Since PCR assays cannot differentiate between an infection and asymptomatic colonization of C. diff, further problems result in those with ulcerative colitis (Reinink et al., 2017). There is a higher prevalence of C. diff colonization in those with ulcerative colitis (Reinink et al., 2017). But symptoms of ulcerative colitis and CDI overlap. If individuals with ulcerative colitis have C. diff colonization that is asymptomatic, they may be confused as having CDI and be treated with antibiotics, when they actually require immunosuppression to treat their colitis (Reinink et al., 2017).

One solution to better diagnostic measures may be searching for other molecules and testing whether the amounts are different in those with ulcerative colitis and those with CDI. This is the approach Reinink et al. (2017) took to attempt to differentiate between the two conditions. They measured procalcitonin (PCT), which serves as a biomarker for other bacterial infections. They hypothesized that PCT would be elevated in acute CDI but not in an ulcerative colitis flare or C. diff colonization. Unfortunately, they found that PCT did not help discriminate acute ulcerative colitis flares from CDI. However, PCT may help discriminate patients who respond to treatment (those that truly have CDI) from those that only have asymptomatic colonization of C. diff (Reinink et al., 2017).

At present, we do not have a good way of appropriately differentiating ulcerative colitis flares from CDI. Since the treatment methods for each type of condition are unique to each other and in some cases even contraindicatory, this presents a problem. One potential solution to the treatment approach could be to discover a treatment that is safe and applicable to both ulcerative colitis patients and those with CDI. As has been extensively discussed in previous chapters, fecal microbiota transplantation (FMT) has been used and currently is being used, to treat recurrent CDI successfully (Narula et al., 2017). Narula et al. (2017) conducted a meta-analysis examining FMT for the treatment of active ulcerative colitis. Based on the results, they concluded that FMT may be beneficial and safe for treating active ulcerative colitis and inducing remission.

Although we may not have appropriate tools for differentiating ulcerative colitis flares from CDI, we could potentially have a treatment that safely addresses both. If FMT could be used in both cases, those with ulcerative colitis would not face all the negative consequences of having an undetected CDI (because even if the CDI is not diagnosed, the FMT used for treating the ulcerative colitis may unintentionally treat the CDI anyways).

Crohn's disease

Crohn's disease is another type of inflammatory bowel disease (IBD) which is associated with inflammation in the digestive tract. Crohn's disease has similar symptoms to other IBD's like colitis, which includes stomach pain and severe diarrhea. Crohn's disease is hypothesized to be a result of immune responses to fecal microorganisms in biologically susceptible individuals (Suskind et al., 2015). Such immune responses could alter the microorganisms, resulting in imbalances that promote further inflammatory responses (Suskind et al., 2015). These imbalances are characterized by depletion of commensal bacteria (bacteria that act on an individual's immune system to induce protective responses against pathogens) (Khan et al., 2019). Several clinical studies have attempted to regulate the fecal microorganisms to decrease inflammatory immune responses using prebiotic, probiotic, and microbial therapies but the results of these studies are mixed (Suskind et al., 2015).

Modifying the human microbiome (the microorganisms in a specific environment, such as in the digestive tract) through FMT may be helpful in those with Crohn's disease (Suskind et al., 2015). The goal of FMT in Crohn's disease is to alter the fecal microbiome by reducing bacteria that may cause imbalances that lead to inflammation (Suskind et al., 2015). A study in a small group of pediatric patients (between twelve and nineteen years old) with Crohn's disease found that FMT was safe and well-tolerated (Suskind et al., 2015). The researchers also found that clinical and laboratory improvements were seen in the majority of patients. The results provide evidence that fecal microorganisms may play an important role in the development of Crohn's disease and indicates that FMT could potentially be used as a therapy for Crohn's disease (Suskind et al., 2015).

Arthritis

Rheumatoid arthritis is a chronic inflammatory autoimmune disease (a condition due to which an individual's immune system inappropriately attacks their cells and tissues). Rheumatoid arthritis is characterized by pain in five or more joints and can potentially lead to joint destruction and disability (Zeng et al., 2020). Genetic and environmental factors are involved in the development of rheumatoid arthritis (Zeng et al., 2020). Part of the environmental factors associated with causing rheumatoid arthritis include microorganisms in the body and infections (Zeng et al., 2020). Due to these factors, antimicrobial drugs have been reported to be effective in some rheumatoid arthritis patients (Zeng et al., 2020). These findings suggest that microorganisms in the stomach may be correlated with rheumatoid arthritis as microorganisms in the gastrointestinal tract play a major role in overall immunity (Zeng et al., 2020).

Some studies have reported that the composition of intestinal microorganisms is altered in rheumatoid arthritis patients, which indicates a potential location for targeting therapy in rheumatoid arthritis (Zeng et al., 2020). All of this evidence also suggests that microorganism imbalances in the stomach play a crucial role in the cause and development of Rheumatoid arthritis (Zeng et al., 2020). In a single case study of a twenty-year-old woman with rheumatoid arthritis, many of the medications she was given to address her joint pain and swelling were not working for her (Zeng et al., 2020). This patient was then treated with FMT, to which she successfully responded. The authors suggest they were able to successfully treat the patient's rheumatoid arthritis with FMT by reconstructing helpful microorganisms in her intestinal tract, which then helped reduce or stop inflammatory processes (Zeng et al., 2020). Thus, FMT may in the future help treat rheumatoid arthritis. It is important to note, however, that this was a single case study. Before any further applications of FMT, these findings must be replicated in an appropriate clinical trial to confirm that the patient's recovery was truly due to FMT and not other external factors. Only after confirming this can FMT become implemented on a larger scale to treat rheumatoid arthritis.

Multiple sclerosis

Multiple sclerosis is an inflammatory, degenerative autoimmune disease that affects the central nervous system (CNS) (Schepici et al., 2019). Similar to many illnesses, genetic and environmental factors play a key role in the development of multiple sclerosis (Schepici et al., 2019). Commensal microorganisms may play an important role in immune-related diseases such as multiple sclerosis (Schepici et al., 2019). Commensal microorganisms are found in the stomach, where they appear to play an important role in the development of multiple sclerosis (Schepici et al., 2019). These organisms appear to be involved in regulating the immune system, altering the structure and function of the blood-brain barrier (protects the brain from harmful molecules and pathogens), triggering autoimmune demyelination (immune system attacks myelin which is insulating material on nerve cells that allows quick signal transmission) and interacts with other cells in the CNS (Schepici et al., 2019). Several studies confirm that a characteristic imbalance in the intestines is constantly present over the course of multiple sclerosis, suggesting a connection between the worsening of multiple sclerosis and commensal bacteria (Schepici et al., 2019).

There are several isolated case studies of patients with multiple sclerosis experiencing a reduction of symptoms after FMT (Borody et al., 2011). In three individual cases, patients with multiple sclerosis received FMT to treat their constipation (Borody et al., 2011). The first case involves a thirty-year-old male who also had symptoms like vertigo, impaired concentration, and severe leg weakness (Borody et al., 2011). This patient required a wheelchair and had a urinary catheter. After five FMT transfusions, the patient's constipation was treated. Interestingly, the patient's multiple sclerosis also improved progressively, they regained the ability to walk, and the catheter was removed. This patient is well fifteen years post-FMT. The other two cases involved a twenty-nine-year-old patient and an eighty-year-old patient (Borody et al., 2011). Both patients' constipation was resolved after FMT transfusions, and they also had improved neurological symptoms and motor ability. Based on these results, Borody et al. (2011) speculate that FMT can reverse multiple sclerosis symptoms and suggest a gastrointestinal infection may underpin these types of disorders.

Since the above-mentioned studies were only case studies, the results must be replicated in clinical trials. At present, there are clinical trials underway evaluating the effectiveness of FMT in treating multiple sclerosis symptoms (Schepici et al., 2019). The results of these trials will guide the future use of FMT, whether it is safe, and how FMT works for multiple sclerosis (Schepici et al., 2019). Thus, FMT may be an effective future treatment for diseases associated with such microorganism alterations, which includes multiple sclerosis (Schepici et al., 2019).

Summary

At present, there are many potential applications for FMT that require further research and clinical trials to confirm their effectiveness. Many of the diseases that could benefit from fecal transplants are those concerning the immune system and inflammation, such as autoimmune diseases (which specifically involve inflammation due to the body's maladaptive responses). Several case studies have been conducted to examine the effectiveness of FMT for treating disorders that are associated with altered organisms in the intestines (Choi & Cho, 2016). There are lots of disorders that fall under this category, including and not limited to: IBD,IBS, colorectal cancer, arthritis, asthma, autism (refer to chapter 6 for more details), chronic fatigue syndrome, diabetes mellitus, eczema, fibromyalgia, hay fever, ischemic heart disease, mood disorders (including anxiety and depressive disorders (refer to chapter 4 and 6 for more information), multiple sclerosis, nonalcoholic fatty liver disease, obesity, and Parkinson's disease (Choi & Cho, 2016). Should clinical trials provide evidence to support FMT's effectiveness in treating these disorders, all of these disorders have the potential to be treated with FMT in the future.

References
Chapter 1.

Villines, Z. (2019). What is a fecal transplant? Everything you need to know. Medical News Today. https://www.medicalnewstoday.com/articles/325128#what-is-a-fecal-transplant

Khoruts A. (2017). Fecal microbiota transplantation-early steps on a long journey ahead. Gut microbes, 8(3), 199–204. https://doi.org/10.1080/19490976.2017.1316447

Digestive System Bacteria. (2018). CK-12. https://flexbooks.ck12.org/cbook/ck-12-middle-school-life-science-2.0/section/11.19/primary/lesson/bacteria-in-the-digestive-system-ms-ls

The Structure and Function of the Digestive System. (2018). Cleveland Clinic. https://my.clevelandclinic.org/health/articles/7041-the-structure-and-function-of-the-digestive-system

De Groot, P. F., Frissen, M. N., de Clercq, N. C., Nieuwdorp, M. (2017). Fecal Microbiota Transplantation in Metabolic Syndrome: History, Present and Future. Gut microbes. https://www.ncbi.nlm.nih.gov/pmc/articles/PMC5479392/#:~:text=The%20history%20of%20fecal%20microbiota,and%20its%20resident%20microbial%20communities

Chapter 2.

Aron-Wisnewsky, J., Clément, K., & Nieuwdorp, M. (2019). Fecal Microbiota Transplantation: A Future Therapeutic Option for Obesity/Diabetes? Current Diabetes Reports, 19(8), 51. https://doi.org/10.1007/s11892-019-1180-z

Azad, M. B., & Kozyrskyj, A. L. (2012). Perinatal programming of asthma: The role of gut microbiota. Clinical & Developmental Immunology, 2012, 932072. https://doi.org/10.1155/2012/932072

Bilinski, J., Grzesiowski, P., Sorensen, N., Madry, K., Muszynski, J., Robak, K., Wroblewska, M., Dzieciatkowski, T., Dulny, G., Dwilewicz-Trojaczek, J., Wiktor-Jedrzejczak, W., & Basak, G. W. (2017). Fecal Microbiota Transplantation in Patients With Blood Disorders Inhibits Gut Colonization With Antibiotic-Resistant Bacteria: Results of a

Prospective, Single-Center Study. Clinical Infectious Diseases, 65(3), 364–370. https://doi.org/10.1093/cid/cix252

Brunt, E. M. (2001). Nonalcoholic Steatohepatitis: Definition and Pathology. Seminars in Liver Disease, 21(1), 3–16. https://doi.org/10.1055/s-2001-12925

Daloiso, V., Minacori, R., Refolo, P., Sacchini, D., Craxì, L., Gasbarrini, A., & Spagnolo, A. G. (2015). Ethical aspects of Fecal Microbiota Transplantation (FMT). European Review for Medical and Pharmacological Sciences, 19(17), 3173–3180.

de Clercq, N. C., Frissen, M. N., Davids, M., Groen, A. K., & Nieuwdorp, M. (2019). Weight Gain after Fecal Microbiota Transplantation in a Patient with Recurrent Underweight following Clinical Recovery from Anorexia Nervosa. Psychotherapy and Psychosomatics, 88(1), 58–60. https://doi.org/10.1159/000495044

Donia, M. S., Cimermancic, P., Schulze, C. J., Wieland Brown, L. C., Martin, J., Mitreva, M., Clardy, J., Linington, R. G., & Fischbach, M. A. (2014). A systematic analysis of biosynthetic gene clusters in the human microbiome reveals a common family of antibiotics. Cell, 158(6), 1402–1414. https://doi.org/10.1016/j.cell.2014.08.032

Dzidic, M., Abrahamsson, T. R., Artacho, A., Björkstén, B., Collado, M. C., Mira, A., & Jenmalm, M. C. (2017). Aberrant IgA responses to the gut microbiota during infancy precede asthma and allergy development. Journal of Allergy and Clinical Immunology, 139(3), 1017-1025.e14. https://doi.org/10.1016/j.jaci.2016.06.047

Eckel, R. H., Alberti, K., Grundy, S. M., & Zimmet, P. Z. (2010). The metabolic syndrome. The Lancet, 375(9710), 181–183. https://doi.org/10.1016/S0140-6736(09)61794-3

Everard, A., & Cani, P. D. (2013). Diabetes, obesity and gut microbiota. Best Practice & Research Clinical Gastroenterology, 27(1), 73–83. https://doi.org/10.1016/j.bpg.2013.03.007

Field, C. J. (2005). The immunological components of human milk and their effect on immune development in infants. The Journal of Nutrition, 135(1), 1–4. https://doi.org/10.1093/jn/135.1.1

Golfeyz. (2018). Celiac Disease and Fecal Microbiota Transplantation: A New Beginning? The American Journal of Gastroenterology; New York, 113(8), 1256. https://doi.org/10.1038/s41395-018-0094-8

Gough, E., Shaikh, H., & Manges, A. R. (2011). Systematic review of intestinal microbiota transplantation (fecal bacteriotherapy) for recurrent clostridium difficile infection. Clinical Infectious Diseases, 53(10), 994–1002. https://doi.org/10.1093/cid/cir632

Grigoryan, Z., Shen, M. J., Twardus, S. W., Beuttler, M. M., Chen, L. A., & Bateman-House, A. (2020). Fecal microbiota transplantation: Uses, questions, and ethics. Medicine in Microecology, 6, 100027. https://doi.org/10.1016/j.medmic.2020.100027

Ianiro, G., Masucci, L., Quaranta, G., Simonelli, C., Lopetuso, L. R., Sanguinetti, M., Gasbarrini, A., & Cammarota, G. (2018). Randomised clinical trial: Faecal microbiota transplantation by colonoscopy plus vancomycin for the treatment of severe refractory Clostridium difficile infection—single versus multiple infusions. Alimentary Pharmacology & Therapeutics, 48(2), 152–159. https://doi.org/10.1111/apt.14816

Kelly, J. R., Minuto, C., Cryan, J. F., Clarke, G., & Dinan, T. G. (2017). Cross Talk: The Microbiota and Neurodevelopmental Disorders. Frontiers in Neuroscience, 11. https://doi.org/10.3389/fnins.2017.00490

Khoruts, A., Sadowsky, M. J., & Hamilton, M. J. (2015). Development of Fecal Microbiota Transplantation Suitable for Mainstream Medicine. Clinical Gastroenterology and Hepatology, 13(2), 246–250. https://doi.org/10.1016/j.cgh.2014.11.014

Li, M., Wang, M., & Donovan, S. M. (2014). Early Development of the Gut Microbiome and Immune-Mediated Childhood Disorders. Seminars in Reproductive Medicine, 32(1), 74–86. https://doi.org/10.1055/s-0033-1361825

Marsot, A., Boulamery, A., Bruguerolle, B., & Simon, N. (2012). Vancomycin. Clinical Pharmacokinetics, 51(1), 1–13. https://doi.org/10.2165/11596390-000000000-00000

McGuire, A. L., Colgrove, J., Whitney, S. N., Diaz, C. M., Bustillos, D., & Versalovic, J. (2008). Ethical, legal, and social considerations in conducting the Human Microbiome Project. Genome Research, 18(12), 1861–1864. https://doi.org/10.1101/gr.081653.108

Pinn, D. M., Aroniadis, O. C., & Brandt, L. J. (2015). Is fecal microbiota transplantation (FMT) an effective treatment for patients with functional gastrointestinal disorders (FGID)? Neurogastroenterology & Motility, 27(1), 19–29. https://doi. org/10.1111/nmo.12479

Rautava, S., & Isolauri, E. (2002). The development of gut immune responses and gut microbiota: Effects of probiotics in prevention and treatment of allergic disease. Current Issues in Intestinal Microbiology, 3(1), 15–22.

Richards, J. L., Yap, Y. A., McLeod, K. H., Mackay, C. R., & Mariño, E. (2016). Dietary metabolites and the gut microbiota: An alternative approach to control inflammatory and autoimmune diseases. Clinical & Translational Immunology, 5(5), e82. https://doi.org/10.1038/cti.2016.29

Sullivan, A., Edlund, C., & Nord, C. E. (2001). Effect of antimicrobial agents on the ecological balance of human microflora. The Lancet. Infectious Diseases, 1(2), 101–114. https://doi.org/10.1016/S1473-3099(01)00066-4

Turnbaugh, P. J., Ley, R. E., Mahowald, M. A., Magrini, V., Mardis, E. R., & Gordon, J. I. (2006). An obesity-associated gut microbiome with increased capacity for energy harvest. Nature, 444(7122), 1027–1031. https://doi.org/10.1038/nature05414

van Nood, E., Speelman, P., Nieuwdorp, M., & Keller, J. (2014). Fecal microbiota transplantation: Facts and controversies. Current Opinion in Gastroenterology, 30(1), 34–39. https://doi.org/10.1097/MOG.0000000000000024

Verbeke, K. A., Boesmans, L., & Boets, E. (2014). Modulating the microbiota in inflammatory bowel diseases: Prebiotics, probiotics or faecal transplantation? Proceedings of the Nutrition Society, 73(4), 490–497. https://doi.org/10.1017/S0029665114000639

Vyas, D., Aekka, A., & Vyas, A. (2015). Fecal transplant policy and legislation. World Journal of Gastroenterology : WJG, 21(1), 6–11.

https://doi.org/10.3748/wjg.v21.i1.6

Wu, R.-Q., Zhang, D.-F., Tu, E., Chen, Q.-M., & Chen, W. (2014). The mucosal immune system in the oral cavity—An orchestra of T cell diversity. International Journal of Oral Science, 6(3), 125–132. https://doi.org/10.1038/ijos.2014.48

Chapter 3.

Braak, H., Rüb, U., Gai, W. P., & Del Tredici, K. (2003). Idiopathic Parkinson's disease: Possible routes by which vulnerable neuronal types may be subject to neuroinvasion by an unknown pathogen. Journal of Neural Transmission, 110(5), 517–536. https://doi.org/10.1007/s00702-002-0808-2

CDC. (2021, March 8). Could you or your loved one have C. diff? Centers for Disease Control and Prevention. https://www.cdc.gov/cdiff/what-is.html

Chaidez, V., Hansen, R. L., & Hertz-Picciotto, I. (2014). Gastrointestinal Problems in Children with Autism, Developmental Delays or Typical Development. Journal of Autism and Developmental Disorders, 44(5), 1117–1127. https://doi.org/10.1007/s10803-013-1973-x

Choi, H. H., & Cho, Y.-S. (2016). Fecal Microbiota Transplantation: Current Applications, Effectiveness, and Future Perspectives. Clinical Endoscopy, 49(3), 257–265. https://doi.org/10.5946/ce.2015.117

Colman, R. J., & Rubin, D. T. (2014). Fecal microbiota transplantation as therapy for inflammatory bowel disease: A systematic review and meta-analysis. Journal of Crohn's & Colitis, 8(12), 1569–1581. https://doi.org/10.1016/j.crohns.2014.08.006

Devos, D., Lebouvier, T., Lardeux, B., Biraud, M., Rouaud, T., Pouclet, H., Coron, E., Bruley des Varannes, S., Naveilhan, P., Nguyen, J.-M., Neunlist, M., & Derkinderen, P. (2013). Colonic inflammation in Parkinson's disease. Neurobiology of Disease, 50, 42–48. https://doi.org/10.1016/j.nbd.2012.09.007

Frank, D. N., St Amand, A. L., Feldman, R. A., Boedeker, E. C., Harpaz, N., & Pace, N. R. (2007). Molecular-phylogenetic characterization of microbial community imbalances in human inflammatory bowel diseases. Proceedings of the National Academy of Sciences of the United States of America, 104(34), 13780–13785. https://doi.org/10.1073/pnas.0706625104

Goloshchapov O.V et al. (2019). Long-term impact of fecal transplantation in healthy volunteers. BMC Microbiology. 19(312). https://doi.org/10.1186/s12866-019-1689-y

Hasegawa, S., Goto, S., Tsuji, H., Okuno, T., Asahara, T., Nomoto, K., Shibata, A., Fujisawa, Y., Minato, T., Okamoto, A., Ohno, K., & Hirayama, M. (2015). Intestinal Dysbiosis and Lowered Serum Lipopolysaccharide-Binding Protein in Parkinson's Disease. PLOS ONE, 10(11), e0142164. https://doi.org/10.1371/journal.pone.0142164
Hooper, L. V., Littman, D. R., & Macpherson, A. J. (2012). Interactions between the microbiota and the immune system. Science (New York, N.Y.), 336(6086), 1268–1273. https://doi.org/10.1126/science.1223490

Jenner, P. (2008). Molecular mechanisms of L-DOPA-induced dyskinesia. Nature Reviews Neuroscience, 9(9), 665–677. https://doi.org/10.1038/nrn2471

Kang, D.-W., Adams, J. B., Gregory, A. C., Borody, T., Chittick, L., Fasano, A., Khoruts, A., Geis, E., Maldonado, J., McDonough-Means, S., Pollard, E. L., Roux, S., Sadowsky, M. J., Lipson, K. S., Sullivan, M. B., Caporaso, J. G., & Krajmalnik-Brown, R. (2017). Microbiota Transfer Therapy alters gut ecosystem and improves gastrointestinal and autism symptoms: An open-label study. Microbiome, 5(1), 10. https://doi.org/10.1186/s40168-016-0225-7

Kang, D.-W., Park, J. G., Ilhan, Z. E., Wallstrom, G., Labaer, J., Adams, J. B., & Krajmalnik-Brown, R. (2013). Reduced incidence of Prevotella and other fermenters in intestinal microflora of autistic children. PloS One, 8(7), e68322. https://doi.org/10.1371/journal.pone.0068322

Niehus, R., & Lord, C. (2006). Early medical history of children with autism spectrum disorders. Journal of Developmental and Behavioral Pediatrics: JDBP, 27(2 Suppl), S120-127. https://doi.org/10.1097/00004703-200604002-00010

Sampson, T. R., Debelius, J. W., Thron, T., Janssen, S., Shastri, G.

G., Ilhan, Z. E., Challis, C., Schretter, C. E., Rocha, S., Gradinaru, V., Chesselet, M.-F., Keshavarzian, A., Shannon, K. M., Krajmalnik-Brown, R., Wittung-Stafshede, P., Knight, R., & Mazmanian, S. K. (2016). Gut Microbiota Regulate Motor Deficits and Neuroinflammation in a Model of Parkinson's Disease. Cell, 167(6), 1469-1480.e12. https://doi.org/10.1016/j.cell.2016.11.018

Svensson, E., Horváth-Puhó, E., Thomsen, R. W., Djurhuus, J. C., Pedersen, L., Borghammer, P., & Sørensen, H. T. (2015). Vagotomy and subsequent risk of Parkinson's disease. Annals of Neurology, 78(4), 522–529. https://doi.org/10.1002/ana.24448

The microbiome, fecal microbiota transplants and inflammatory bowel disease—Mayo Clinic. (n.d.). Retrieved May 16, 2021, from https://www.mayoclinic.org/medical-professionals/digestive-diseases/news/the-microbiome-fecal-microbiota-transplants-and-inflammatory-bowel-disease/mqc-20463208

Villines Z. (2019). What is a fecal transplant? Everything you need to know. https://www.medicalnewstoday.com/articles/325128

Xu, M.-Q., Cao, H.-L., Wang, W.-Q., Wang, S., Cao, X.-C., Yan, F., & Wang, B.-M. (2015). Fecal microbiota transplantation broadening its application beyond intestinal disorders. World Journal of Gastroenterology, 21(1), 102–111. https://doi.org/10.3748/wjg.v21.i1.102

Youngster, I., Sauk, J., Pindar, C., Wilson, R. G., Kaplan, J. L., Smith, M. B., Alm, E. J., Gevers, D., Russell, G. H., & Hohmann, E. L. (2014). Fecal Microbiota Transplant for Relapsing Clostridium difficile Infection Using a Frozen Inoculum From Unrelated Donors: A Randomized, Open-Label, Controlled Pilot Study. Clinical Infectious Diseases: An Official Publication of the Infectious Diseases Society of America, 58(11), 1515–1522. https://doi.org/10.1093/cid/ciu135

Chapter 4.

Carabotti, M., Scirocco, A., Maselli, M. A., & Severi, C. (2015). The gut-brain axis: interactions between enteric microbiota, central and enteric nervous systems. Annals of gastroenterology, 28(2), 203–209.

Chinna Meyyappan, A., Forth, E., Wallace, C., & Milev, R. (2020).

Effect of fecal microbiota transplant on symptoms of psychiatric disorders: a systematic review. BMC psychiatry, 20(1), 299. https://doi.org/10.1186/s12888-020-02654-5

Choi, H. H., & Cho, Y. S. (2016). Fecal Microbiota Transplantation: Current Applications, Effectiveness, and Future Perspectives. Clinical endoscopy, 49(3), 257–265. https://doi.org/10.5946/ce.2015.117

Gutin, L., Piceno, Y., Fadrosh, D., Lynch, K., Zydek, M., Kassam, Z., LaMere, B., Terdiman, J., Ma, A., Somsouk, M., Lynch, S., & El-Nachef, N. (2019). Fecal microbiota transplant for Crohn's disease: A study evaluating safety, efficacy, and microbiome profile. United European gastroenterology journal, 7(6), 807–814. https://doi.org/10.1177/2050640619845986

Jiang, C., Li, G., Huang, P., Liu, Z., & Zhao, B. (2017). The Gut Microbiota and Alzheimer's Disease. Journal of Alzheimer's disease : JAD, 58(1), 1–15. https://doi.org/10.3233/JAD-161141

Konturek, P. C., Koziel, J., Dieterich, W., Haziri, D., Wirtz, S., Glowczyk, I., Konturek, K., Neurath, M. F., & Zopf, Y. (2016). Successful therapy of Clostridium difficile infection with fecal microbiota transplantation. Journal of physiology and pharmacology : an official journal of the Polish Physiological Society, 67(6), 859–866.

Lange, K., Buerger, M., Stallmach, A., & Bruns, T. (2016). Effects of Antibiotics on Gut Microbiota. Digestive diseases (Basel, Switzerland), 34(3), 260–268. https://doi.org/10.1159/000443360

Lee, P., Yacyshyn, B. R., & Yacyshyn, M. B. (2019). Gut microbiota and obesity: An opportunity to alter obesity through fecal microbiota transplant (FMT). Diabetes, obesity & metabolism, 21(3), 479–490. https://doi.org/10.1111/dom.13561

Marotz, C. A., & Zarrinpar, A. (2016). Treating Obesity and Metabolic Syndrome with Fecal Microbiota Transplantation. The Yale journal of biology and medicine, 89(3), 383–388.

Meighani, A., Hart, B. R., Mittal, C., Miller, N., John, A., & Ramesh, M. (2016). Predictors of fecal transplant failure. European journal of gastroenterology & hepatology, 28(7), 826–830. https://doi.org/10.1097/MEG.0000000000000614

Mullish, B. H., & Williams, H. R. (2018). Clostridium difficile infection and antibiotic-associated diarrhoea. Clinical medicine (London, England), 18(3), 237–241. https://doi.org/10.7861/clinmedicine.18-3-237

Wang, A. Y., Popov, J., & Pai, N. (2016). Fecal microbial transplant for the treatment of pediatric inflammatory bowel disease. World journal of gastroenterology, 22(47), 10304–10315.https://doi.org/10.3748/wjg.v22.i47.10304

Weingarden, A. R., & Vaughn, B. P. (2017). Intestinal microbiota, fecal microbiota transplantation, and inflammatory bowel disease. Gut microbes, 8(3), 238–252.https://doi.org/10.1080/19490976.2017.1290757

Youngster, I., Sauk, J., Pindar, C., Wilson, R. G., Kaplan, J. L., Smith, M. B., Alm, E. J., Gevers, D., Russell, G. H., & Hohmann, E. L. (2014). Fecal microbiota transplant for relapsing Clostridium difficile infection using a frozen inoculum from unrelated donors: a randomized, open-label, controlled pilot study. Clinical infectious diseases : an official publication of the Infectious Diseases Society of America, 58(11), 1515–1522. https://doi.org/10.1093/cid/ciu135

Chapter 5.

Allegretti, J. R., Korzenik, J. R., & Hamilton, M. J. (2014). Fecal microbiota transplantation via colonoscopy for recurrent C. difficile Infection. Journal of visualized experiments : JoVE, (94), 52154. https://doi.org/10.3791/52154

Bakken, J. S., Borody, T., Brandt, L. J., Brill, J. V., Demarco, D. C., Franzos, M. A., Kelly, C., Khoruts, A., Louie, T., Martinelli, L. P., Moore, T. A., Russell, G., Surawicz, C., & Fecal Microbiota Transplantation Workgroup. (2011). Treating Clostridium difficile infection with fecal microbiota transplantation. Clinical gastroenterology and hepatology : the official clinical practice journal of the American Gastroenterological Association, 9(12), 1044–1049. https://doi.org/10.1016/j.cgh.2011.08.014

Bibbò, S., Settanni, C. R., Porcari, S., Bocchino, E., Ianiro, G., Cammarota, G., & Gasbarrini, A. (2020). Fecal Microbiota Transplantation: Screening and Selection to Choose the Optimal

Donor. Journal of clinical medicine, 9(6), 1757. https://doi.
org/10.3390/jcm9061757

Boston Children's Hospital. (n.d.). Fecal Microbiota Transplant (FMT).
https://www.childrenshospital.org/conditions-and-treatments/
treatments/fmt

Cammarota, G., Ianiro, G., Tilg, H., Rajilić-Stojanović, M., Kump, P.,
Satokari, R., Sokol, H., Arkkila, P., Pintus, C., Hart, A., Segal, J., Aloi,
M., Masucci, L., Molinaro, A., Scaldaferri, F., Gasbarrini, G., Lopez-
Sanroman, A., Link, A., de Groot, P., de Vos, W. M., … European
FMT Working Group (2017). European consensus conference on
faecal microbiota transplantation in clinical practice. Gut, 66(4),
569–580. https://doi.org/10.1136/gutjnl-2016-313017

Feeding Tube Awareness Foundation. (n.d.). Nasal Tubes (NG, ND,
NJ). https://www.feedingtubeawareness.org/nasal-tubes/

Goldenberg, S. D., Batra, R., Beales, I., Digby-Bell, J. L., Irving, P.
M., Kellingray, L., Narbad, A., & Franslem-Elumogo, N. (2018).
Comparison of Different Strategies for Providing Fecal Microbiota
Transplantation to Treat Patients with Recurrent Clostridium difficile
Infection in Two English Hospitals: A Review. Infectious diseases and
therapy, 7(1), 71–86. https://doi.org/10.1007/s40121-018-0189-y

Healthwise. (n.d.). Learning About a Fecal Transplant. MyHealth.
Alberta.ca. https://myhealth.alberta.ca/Health/aftercareinformation/
pages/conditions.aspx?HwId=acg4186

Jørgensen, S. M. , Hansen, M. M. , Erikstrup, C. , Dahlerup, J. F.
& Hvas, C. L. (2017). Faecal microbiota transplantation. European
Journal of Gastroenterology & Hepatology, 29(11), e36–e45. doi:
10.1097/MEG.0000000000000958.

Kassam, Z. (2018). Clinical Primer: Position Statement for Fecal
Microbiota Transplantation Administration for Recurrent Clostridium
difficile Infection [Position Statement]. OpenBiome. https://static1.
squarespace.com/static/50e0c29ae4b0a05702af7e6a/t/5b7b261503ce6
46d8c7ce0f4/1534797341261/Clinical+Primer.pdf

Kao, D., Roach, B., Silva, M., Beck, P., Rioux, K., Kaplan, G. G.,
Chang, H. J., Coward, S., Goodman, K. J., Xu, H., Madsen, K.,

Mason, A., Wong, G. K., Jovel, J., Patterson, J., & Louie, T. (2017). Effect of Oral Capsule- vs Colonoscopy-Delivered Fecal Microbiota Transplantation on Recurrent Clostridium difficile Infection: A Randomized Clinical Trial. JAMA, 318(20), 1985–1993. https://doi.org/10.1001/jama.2017.17077

Kelly, C. R., Kahn, S., Kashyap, P., Laine, L., Rubin, D., Atreja, A., Moore, T., & Wu, G. (2015). Update on Fecal Microbiota Transplantation 2015: Indications, Methodologies, Mechanisms, and Outlook. Gastroenterology, 149(1), 223–237. https://doi.org/10.1053/j.gastro.2015.05.008

Kump, P. K., Krause, R., Allerberger, F., & Högenauer, C. (2014). Faecal microbiota transplantation—the Austrian approach. Clinical microbiology and infection : the official publication of the European Society of Clinical Microbiology and Infectious Diseases, 20(11), 1106–1111. https://doi.org/10.1111/1469-0691.12801

Merenstein, D., El-Nachef, N., & Lynch, S. V. (2014). Fecal microbial therapy: promises and pitfalls. Journal of pediatric gastroenterology and nutrition, 59(2), 157–161. https://doi.org/10.1097/MPG.0000000000000415

Mullish, B. H., Quraishi, M. N., Segal, J. P., McCune, V. L., Baxter, M., Marsden, G. L., Moore, D. J., Colville, A., Bhala, N., Iqbal, T. H., Settle, C., Kontkowski, G., Hart, A. L., Hawkey, P. M., Goldenberg, S. D., & Williams, H. (2018). The use of faecal microbiota transplant as treatment for recurrent or refractory Clostridium difficile infection and other potential indications: joint British Society of Gastroenterology (BSG) and Healthcare Infection Society (HIS) guidelines. Journal of Hospital Infection Gut, 100(Supplement 1), S1–S31. https://doi.org/10.1016/j.jhin.2018.07.037

Nicco, C., Paule, A., Konturek, P., & Edeas, M. (2020). From Donor to Patient: Collection, Preparation and Cryopreservation of Fecal Samples for Fecal Microbiota Transplantation. Diseases (Basel, Switzerland), 8(2), 9. https://doi.org/10.3390/diseases8020009

O'Connor, C. (2018, December 4). Everything You Need to Know About Fecal Transplant Pills & Capsules. Designer Shit Documentary. https://designershitdocumentary.com/fmt-capsules-pills/

O'Neill, M. (2019, June 17). How Is a Fecal Transplant Done? Here's Why the FDA is Now Warning Patients About Them. Health. https://www.health.com/condition/digestive-health/how-is-a-fecal-transplant-done

Seladi-Schulman, J. (2019, April 3). Fecal Transplants: The Key To Improving Gut Health? Healthline. https://www.healthline.com/health/fecal-transplants-the-key-to-improving-gut-health#procedure

Villines, Z. (2019, May 8). Everything you need to know about fecal transplants. Medical News Today. https://www.medicalnewstoday.com/articles/325128

Whitlock, J. (2020, April 5). Types of Feeding Tubes and Their Uses. Verywell Health. https://www.verywellhealth.com/what-you-should-know-about-feeding-tubes-4152086

Chapter 6.

Baruch, E. N., Youngster, I., Ben-Betzalel, G., Ortenberg, R., Lahat, A., Katz, L., … Boursi, B. (2020). Fecal microbiota transplant promotes response in immunotherapy-refractory melanoma patients. Science, 371(6529), 602–609. https://doi.org/10.1126/science.abb5920

Carabotti, M., Scirocco, A., Maselli, M., & Carola. (2015). The gut-brain axis: interactions between enteric microbiota, central and enteric nervous systems. Annals of Gastroenterology, 28(2), 203–209.

Chinna Meyyappan, A., Forth, E., Wallace, C. J., & Milev, R. (2020). Effect of fecal microbiota transplant on symptoms of psychiatric disorders: a systematic review. BMC Psychiatry, 20(1). https://doi.org/10.1186/s12888-020-02654-5

El-Salhy, M., Hatlebakk, J. G., Gilja, O. H., Bråthen Kristoffersen, A., & Hausken, T. (2019). Efficacy of faecal microbiota transplantation for patients with irritable bowel syndrome in a randomised, double-blind, placebo-controlled study. Gut, 69(5), 859–867. https://doi.org/10.1136/gutjnl-2019-319630

Graefe, S., & Mohiuddin, S. S. (2020, May 7). Biochemistry, Substance P. StatPearls [Internet]. https://www.ncbi.nlm.nih.gov/books/NBK554583/.

Gutin, L., Piceno, Y., Fadrosh, D., Lynch, K., Zydek, M., Kassam, Z., ... El-Nachef, N. (2019). Fecal microbiota transplant for Crohn disease: A study evaluating safety, efficacy, and microbiome profile. United European Gastroenterology Journal, 7(6), 807–814. https://doi.org/10.1177/2050640619845986

Harvard School of Public Health. (2020, May 1). The Microbiome. The Nutrition Source. https://www.hsph.harvard.edu/nutritionsource/microbiome/.

Kang, D.-W., Adams, J. B., Coleman, D. M., Pollard, E. L., Maldonado, J., McDonough-Means, S., ... Krajmalnik-Brown, R. (2019). Long-term benefit of Microbiota Transfer Therapy on autism symptoms and gut microbiota. Scientific Reports, 9(1). https://doi.org/10.1038/s41598-019-42183-0

Kang, D.-W., Adams, J. B., Gregory, A. C., Borody, T., Chittick, L., Fasano, A., ... Krajmalnik-Brown, R. (2017). Microbiota Transfer Therapy alters gut ecosystem and improves gastrointestinal and autism symptoms: an open-label study. Microbiome, 5(1). https://doi.org/10.1186/s40168-016-0225-7

National Cancer Institute. (2021, February 4). Fecal transplants help patients respond to immunotherapy. National Cancer Institute. https://www.cancer.gov/news-events/press-releases/2021/fecal-transplants-cancer-immunotherapy.

Rege, S., & Graham, J. (2020, December 3). The simplified guide to the gut-brain axis - how the gut talks to the brain. https://psychscenehub.com/psychinsights/the-simplified-guide-to-the-gut-brain-axis/.

Sah, R., & Geracioti, T. D. (2012). Neuropeptide Y and posttraumatic stress disorder. Molecular Psychiatry, 18(6), 646–655. https://doi.org/10.1038/mp.2012.101

Shreiner, A. B., Kao, J. Y., & Young, V. B. (2015). The gut microbiome in health and in disease. Current Opinion in Gastroenterology, 31(1), 69–75. https://doi.org/10.1097/mog.0000000000000139

Society for Endocrinology. (2018, February). Peptide YY. You and Your

Hormones. https://www.yourhormones.info/hormones/peptide-yy/.

Sokol, H., Landman, C., Seksik, P., Berard, L., Montil, M., Nion-Larmurier, I., … Simon, T. (2020). Fecal microbiota transplantation to maintain remission in Crohn's disease: a pilot randomized controlled study. Microbiome, 8(1). https://doi.org/10.1186/s40168-020-0792-5

Waxenbaum, J. A., Reddy, V., & Varacallo, M. (2020, August 10). Anatomy, autonomic nervous system. https://www.ncbi.nlm.nih.gov/books/NBK539845/.

Zane, T., & Holehan, K. (2018). Focus on science: Is there science behind that?: Fecal microbial transplantation. Science in Autism Treatment, 15(3), 27–29. https://asatonline.org/for-parents/becoming-a-savvy-consumer/is-there-science-behind-that-fecal-microbial-transplantation/.

Chapter 7.

Alang, N., & Kelly, C. R. (2015). Weight Gain After Fecal Microbiota Transplantation. Open Forum Infectious Diseases, 2(1). https://doi.org/10.1093/ofid/ofv004

Bokulich, N. A., Chung, J., Battaglia, T., Henderson, N., Jay, M., Li, H., … Blaser, M. J. (2016). Antibiotics, birth mode, and diet shape microbiome maturation during early life. Science Translational Medicine, 8(343). https://doi.org/10.1126/scitranslmed.aad7121

Borody, T. J., Paramsothy, S., & Agrawal, G. (2013). Fecal Microbiota Transplantation: Indications, Methods, Evidence, and Future Directions. Current Gastroenterology Reports, 15(8). https://doi.org/10.1007/s11894-013-0337-1

Ekekezie, C., Perler, B. K., Wexler, A., Duff, C., Lillis, C. J., & Kelly, C. R. (2020). Understanding the Scope of Do-It-Yourself Fecal Microbiota Transplant. American Journal of Gastroenterology, 115(4), 603–607. https://doi.org/10.14309/ajg.0000000000000499

Fujimura, K. E., Sitarik, A. R., Havstad, S., Lin, D. L., Levan, S., Fadrosh, D., … Lynch, S. V. (2016). Neonatal gut microbiota associates with childhood multisensitized atopy and T cell differentiation. Nature Medicine, 22(10), 1187–1191. https://doi.org/10.1038/nm.4176

Grigoryan, Z., Shen, M. J., Twardus, S. W., Beuttler, M. M., Chen, L. A., & Bateman-House, A. (2020). Fecal microbiota transplantation: Uses, questions, and ethics. Medicine in Microecology, 6, 100027. https://doi.org/10.1016/j.medmic.2020.100027

Gulati, M., Singh, S. K., Corrie, L., Kaur, I. P., & Chandwani, L. (2020). Delivery routes for faecal microbiota transplants: Available, anticipated and aspired. Pharmacological Research, 159, 104954. https://doi.org/10.1016/j.phrs.2020.104954

Koeth, R. A., Wang, Z., Levison, B. S., Buffa, J. A., Org, E., Sheehy, B. T., ... Hazen, S. L. (2013). Intestinal microbiota metabolism of l-carnitine, a nutrient in red meat, promotes atherosclerosis. Nature Medicine, 19(5), 576–585. https://doi.org/10.1038/nm.3145

Kostic, A. D., Gevers, D., Pedamallu, C. S., Michaud, M., Duke, F., Earl, A. M., ... Meyerson, M. (2011). Genomic analysis identifies association of Fusobacterium with colorectal carcinoma. Genome Research, 22(2), 292–298. https://doi.org/10.1101/gr.126573.111

Martinez-Gili, L., McDonald, J. a, Liu, Z., Kao, D., Allegretti, J. R., Monaghan, T. M., ... Mullish, B. H. (2020). Understanding the mechanisms of efficacy of fecal microbiota transplant in treating recurrent Clostridioides difficile infection and beyond: the contribution of gut microbial-derived metabolites. Gut Microbes, 12(1), 1810531. https://doi.org/10.1080/19490976.2020.1810531

Mullish, B. H., Quraishi, M. N., Segal, J. P., McCune, V. L., Baxter, M., Marsden, G. L., ... Goldenberg, S. D. (2018). The use of faecal microbiota transplant as treatment for recurrent or refractory Clostridium difficile infection and other potential indications: joint British Society of Gastroenterology (BSG) and Healthcare Infection Society (HIS) guidelines. Journal of Hospital Infection, 100. https://doi.org/10.1016/j.jhin.2018.07.037

Ott, S. J., Waetzig, G. H., Rehman, A., Moltzau-Anderson, J., Bharti, R., Grasis, J. A., ... Schreiber, S. (2017). Efficacy of Sterile Fecal Filtrate Transfer for Treating Patients With Clostridium difficile Infection. Gastroenterology, 152(4). https://doi.org/10.1053/j.gastro.2016.11.010

Owens, C., Broussard, E., & Surawicz, C. (2013). Fecal microbiota transplantation and donor standardization. Trends in Microbiology, 21(9), 443–445. https://doi.org/10.1016/j.tim.2013.07.003

Schulfer, A. F., Battaglia, T., Alvarez, Y., Bijnens, L., Ruiz, V. E., Ho, M., ... Blaser, M. J. (2017). Intergenerational transfer of antibiotic-perturbed microbiota enhances colitis in susceptible mice. Nature Microbiology, 3(2), 234–242. https://doi.org/10.1038/s41564-017-0075-5

Seekatz, A. M., Theriot, C. M., Rao, K., Chang, Y.-M., Freeman, A. E., Kao, J. Y., & Young, V. B. (2018). Restoration of short chain fatty acid and bile acid metabolism following fecal microbiota transplantation in patients with recurrent Clostridium difficile infection. Anaerobe, 53, 64–73. https://doi.org/10.1016/j.anaerobe.2018.04.001

Smillie, C. S., Sauk, J., Gevers, D., Friedman, J., Sung, J., Youngster, I., ... Alm, E. J. (2018). Strain Tracking Reveals the Determinants of Bacterial Engraftment in the Human Gut Following Fecal Microbiota Transplantation. Cell Host & Microbe, 23(2). https://doi.org/10.1016/j.chom.2018.01.003

Sorg, J. A., & Sonenshein, A. L. (2008). Bile Salts and Glycine as Cogerminants for Clostridium difficile Spores. Journal of Bacteriology, 190(7), 2505–2512. https://doi.org/10.1128/jb.01765-07

Studer, N., Desharnais, L., Beutler, M., Brugiroux, S., Terrazos, M. A., Menin, L., ... Hapfelmeier, S. (2016). Functional Intestinal Bile Acid 7α-Dehydroxylation by Clostridium scindens Associated with Protection from Clostridium difficile Infection in a Gnotobiotic Mouse Model. Frontiers in Cellular and Infection Microbiology, 6. https://doi.org/10.3389/fcimb.2016.00191

Theriot, C. M., Bowman, A. A., & Young, V. B. (2016). Antibiotic-Induced Alterations of the Gut Microbiota Alter Secondary Bile Acid Production and Allow for Clostridium difficile Spore Germination and Outgrowth in the Large Intestine. MSphere, 1(1). https://doi.org/10.1128/msphere.00045-15

Woodworth, M. H., Hayden, M. K., Young, V. B., & Kwon, J. H. (2019). The Role of Fecal Microbiota Transplantation in Reducing Intestinal Colonization With Antibiotic-Resistant Organisms: The

Current Landscape and Future Directions. Open Forum Infectious Diseases, 6(7). https://doi.org/10.1093/ofid/ofz288

Wu, S., Rhee, K.-J., Albesiano, E., Rabizadeh, S., Wu, X., Yen, H.-R., ... Sears, C. L. (2009). A human colonic commensal promotes colon tumorigenesis via activation of T helper type 17 T cell responses. Nature Medicine, 15(9), 1016–1022. https://doi.org/10.1038/nm.2015

Żebrowska, P., Łaczmańska, I., & Łaczmański, Ł. (2021). Future Directions in Reducing Gastrointestinal Disorders in Children With ASD Using Fecal Microbiota Transplantation. Frontiers in Cellular and Infection Microbiology, 11. https://doi.org/10.3389/fcimb.2021.630052

Chapter 8.

Allied Health. (2020, June 23). The vagus nerve: Your secret weapon in fighting stress. Retrieved May 16, 2021, from https://www.allied-services.org/news/2020/june/the-vagus-nerve-your-secret-weapon-in-fighting-s/

Arjomand, A. (2020). P167 adult Crohn's Disease treated with the specific carbohydrate diet (SCD): A case report using objective markers of response. Gastroenterology, 158(3). doi:10.1053/j.gastro.2019.11.164

Bojanova, D. P., & Bordenstein, S. R. (2016). Fecal Transplants: What Is Being Transferred? PLOS Biology, 14(7). doi:10.1371/journal.pbio.1002503

Breit, S., Kupferberg, A., Rogler, G., & Hasler, G. (2018). Vagus Nerve as Modulator of the Brain–Gut Axis in Psychiatric and Inflammatory Disorders. Frontiers in Psychiatry, 9. doi:10.3389/fpsyt.2018.00044

Castro, I., Tasias, M., Calabuig, E., & Salavert, M. (2019). Doctor, my patient has CDI and should continue to receive antibiotics: The (unresolved) risk of recurrent CDI. Revista espanola de quimioterapia : publicacion oficial de la Sociedad Espanola de Quimioterapia, 32 Suppl 2(Suppl 2), 47–54.

Chaudhry, N., & Duggal, A. K. (2014). Sepsis Associated Encephalopathy. Advances in Medicine, 2014, 1-16.

doi:10.1155/2014/762320

Dao, M. C., Everard, A., Aron-Wisnewsky, J., Sokolovska, N., Prifti, E., Verger, E. O., . . . Clément, K. (2015). Akkermansia muciniphila and improved metabolic health during a dietary intervention in obesity: Relationship with gut microbiome richness and ecology. Gut, 65(3), 426-436. doi:10.1136/gutjnl-2014-308778

Dill-McFarland, K. A., Tang, Z., Kemis, J. H., Kerby, R. L., Chen, G., Palloni, A., . . . Herd, P. (2019). Close social relationships correlate with human gut microbiota composition. Scientific Reports, 9(1). doi:10.1038/s41598-018-37298-9

Fonti, R., Latella, G., Bises, G., Magliocca, F., Nobili, F., Caprilli, R., & Sambuy, Y. (1994). Human colonocytes in primary culture: A model to study epithelial growth, metabolism and differentiation. International Journal of Colorectal Disease, 9(1), 13-22. doi:10.1007/bf00304294

Ge, X., Tian, H., Ding, C., Gu, L., Wei, Y., Gong, J., . . . Li, J. (2016). Fecal Microbiota Transplantation in Combination with Soluble Dietary Fiber for Treatment of Slow Transit Constipation: A Pilot Study. Archives of Medical Research, 47(3), 236-242. doi:10.1016/j.arcmed.2016.06.005

Goodrich, J., Waters, J., Poole, A., Sutter, J., Koren, O., Blekhman, R., . . . Ley, R. (2014). Human genetics shape the gut microbiome. Cell, 159(4), 789-799. doi:10.1016/j.cell.2014.09.053

Gough, E., Shaikh, H., & Manges, A. R. (2011). Systematic review of intestinal Microbiota Transplantation (Fecal Bacteriotherapy) for Recurrent Clostridium difficile infection. Clinical Infectious Diseases, 53(10), 994-1002. doi:10.1093/cid/cir632

Howland, R. H. (2014). Vagus Nerve Stimulation. Current Behavioral Neuroscience Reports, 1(2), 64-73. doi:10.1007/s40473-014-0010-5

Kahn, S. A., Vachon, A., Rodriquez, D., Goeppinger, S. R., Surma, B., Marks, J., & Rubin, D. T. (2013). Patient Perceptions of Fecal Microbiota Transplantation for Ulcerative Colitis. Inflammatory Bowel Diseases, 19(7), 1506-1513. doi:10.1097/mib.0b013e318281f520

Kim, Y., & Shin, C. (2018). The Microbiota-Gut-Brain Axis in

Neuropsychiatric Disorders: Pathophysiological Mechanisms and Novel Treatments. Current Neuropharmacology, 16(5), 559-573. doi:10.2174/1570159x15666170915141036

Li, S., Lv, J., Li, J., Zhao, Z., Guo, H., Zhang, Y., . . . Li, Z. (2018). Intestinal microbiota impact sepsis associated encephalopathy via the vagus nerve. Neuroscience Letters, 662, 98-104. doi:10.1016/j.neulet.2017.10.008

Litvak, Y., Byndloss, M. X., & Bäumler, A. J. (2018). Colonocyte metabolism shapes the gut microbiota. Science, 362(6418). doi:10.1126/science.aat9076

Mailing, L. J., Allen, J. M., Buford, T. W., Fields, C. J., & Woods, J. A. (2019). Exercise and the Gut Microbiome: A Review of the Evidence, Potential Mechanisms, and Implications for Human Health. Exercise and Sport Sciences Reviews, 47(2), 75-85. doi:10.1249/jes.0000000000000183

Makki, K., Deehan, E. C., Walter, J., & Bäckhed, F. (2018). The impact of dietary fiber on gut microbiota in host health and disease. Cell Host & Microbe, 23(6), 705-715. doi:10.1016/j.chom.2018.05.012

Meader, E., Mayer, M. J., Steverding, D., Carding, S. R., & Narbad, A. (2013). Evaluation of bacteriophage therapy to control clostridium difficile and toxin production in an in vitro human colon model system. Anaerobe, 22, 25-30. doi:10.1016/j.anaerobe.2013.05.001

Park, L., Mone, A., Price, J. C., Tzimas, D., Hirsh, J., Poles, M. A., & Chen, L. A. (2017). Perceptions of fecal microbiota transplantation for Clostridium difficile infection: Factors that predict acceptance. Annals of Gastroenterology. doi:10.20524/aog.2016.0098

Seeras, K. (2021, January 07). Truncal Vagotomy. Retrieved May 14, 2021, from https://www.ncbi.nlm.nih.gov/books/NBK526104/

Vagus nerve stimulation. (2020, November 17). Retrieved May 16, 2021, from https://www.mayoclinic.org/tests-procedures/vagus-nerve-stimulation/about/pac-20384565

Xu, D., Chen, V. L., Steiner, C. A., Berinstein, J. A., Eswaran, S., Waljee, A. K., . . . Owyang, C. (2019). Efficacy of Fecal Microbiota

Transplantation in Irritable Bowel Syndrome: A Systematic Review and Meta-Analysis. American Journal of Gastroenterology, 114(7), 1043-1050. doi:10.14309/ajg.0000000000000198

Yoon, H., Shim, H. I., Seol, M., Shin, C. M., Park, Y. S., Kim, N., & Lee, D. H. (2021). Factors Related to Outcomes of Fecal Microbiota Transplantation in Patients with Clostridioides difficile Infection. Gut and Liver, 15(1), 61-69. doi:10.5009/gnl20135

Yui, S., Nakamura, T., Sato, T., Nemoto, Y., Mizutani, T., Zheng, X., . . . Watanabe, M. (2012). Functional engraftment of colon epithelium expanded in vitro from a single adult lgr5+ stem cell. Nature Medicine, 18(4), 618-623. doi:10.1038/nm.2695

Zoll, J., Read, M. N., Heywood, S. E., Estevez, E., Marshall, J. P., Kammoun, H. L., . . . Henstridge, D. C. (2020). Fecal microbiota transplantation from high caloric-fed donors alters glucose metabolism in recipient mice, independently of adiposity or exercise status. American Journal of Physiology-Endocrinology and Metabolism, 319(1). doi:10.1152/ajpendo.00037.2020

Chapter 9.

Alagna, L., Haak, B. W., & Gori, A. (2019). Fecal microbiota transplantation in the ICU: perspectives on future implementations. Intensive Care Medicine, 45(7), 998–1001. https://doi.org/10.1007/s00134-019-05645-7

Bilinski, J., Grzesiowski, P., Sorensen, N., Madry, K., Muszynski, J., Robak, K., … Basak, G. W. (2017). Fecal Microbiota Transplantation in Patients With Blood Disorders Inhibits Gut Colonization With Antibiotic-Resistant Bacteria: Results of a Prospective, Single-Center Study. Clinical Infectious Diseases, 65(3), 364–370. https://doi.org/10.1093/cid/cix252

El-Matary, W. (2013). Fecal Microbiota Transplantation: Long-Term Safety Issues. American Journal of Gastroenterology, 108(9), 1537–1538. https://doi.org/10.1038/ajg.2013.208

FDA. (2019). Safety Communication on Use of FMT and MDROs. U.S. Food and Drug Administration. https://www.fda.gov/vaccines-blood-biologics/safety-availability-biologics/important-safety-alert-regarding-

use-fecal-microbiota-transplantation-and-risk-serious-adverse.
Fischer, M., Kao, D., Kelly, C., Kuchipudi, A., Jafri, S.-M.,
Blumenkehl, M., … Allegretti, J. R. (2016). Fecal Microbiota
Transplantation is Safe and Efficacious for Recurrent or Refractory
Clostridium difficile Infection in Patients with Inflammatory Bowel
Disease. Inflammatory Bowel Diseases, 22(10), 2402–2409. https://doi.
org/10.1097/mib.0000000000000908

Glauser, W. (2011). Risk and rewards of fecal transplants. CMAJ.
https://www.cmaj.ca/content/183/5/541.

Hawkins, A. K., & O'Doherty, K. C. (2011). "Who owns your poop?":
insights regarding the intersection of human microbiome research and
the ELSI aspects of biobanking and related studies. BMC Medical
Genomics, 4(1). https://doi.org/10.1186/1755-8794-4-72

Kunde, S., Pham, A., Bonczyk, S., Crumb, T., Duba, M., Conrad, H.,
… Kugathasan, S. (2013). Safety, Tolerability, and Clinical Response
After Fecal Transplantation in Children and Young Adults With
Ulcerative Colitis. Journal of Pediatric Gastroenterology & Nutrition,
56(6), 597–601. https://doi.org/10.1097/mpg.0b013e318292fa0d

Nicholson, M., Thomsen, I., & Edwards, K. (2014). Controversies
Surrounding Clostridium difficile Infection in Infants and
Young Children. Children, 1(1), 40–47. https://doi.org/10.3390/
children1010040

Ma, Y., Liu, J., Rhodes, C., Nie, Y., & Zhang, F. (2017). Ethical Issues
in Fecal Microbiota Transplantation in Practice. The American Journal
of Bioethics, 17(5), 34–45. https://doi.org/10.1080/15265161.2017.1299
240

Nishida, A., Imaeda, H., Bamba, S., & Andoh, A. (2017). Efficacy and
Safety of Single Fecal Microbiota Transplantation for Japanese Patients
with Mild to Moderately Active Ulcerative Colitis. Gastroenterology,
152(5), 476–482. https://doi.org/10.1016/s0016-5085(17)32077-2

Padma. (2019). Does Controversial Poop Transplant Treatment
Have Side-Effects? The Wire. https://thewire.in/the-sciences/does-
controversial-poop-transplant-treatment-have-side-effects.

Paramsothy, M., Borody, T. J., Lin, E., Finlayson, S., Walsh, A.

J., Samuel, D., Bogaerde, V. D., Connor, S., Ng, W., Mitchell, H. M., Kaakoush, N., & A, M.. (2015). Donor Recruitment for Fecal Microbiota Transplantation, Inflammatory Bowel Diseases, 27(7),1600–1606, https://doi.org/10.1097/MIB.0000000000000405

Rubin, Z. A., Martin, E. M., & Allyn, P. (2018). Primary Prevention of Clostridium difficile-Associated Diarrhea: Current Controversies and Future Tools. Current Infectious Disease Reports, 20(9), 1–8. https://doi.org/10.1007/s11908-018-0639-4

Sachs, R. E., & Edelstein, C. A. (2015). Ensuring the safe and effective FDA regulation of fecal microbiota transplantation. Journal of Law and theBiosciences, 2(2), 396–415. https://doi.org/10.1093/jlb/lsv032

Shanahan, F., & Quigley, E. M. (2014). Modification of the gut microbiome to maintain Health or treat disease. Gastroenterology, 146, 1554–1563.

Stallmach, A., Steube, A., Grunert, P., Hartmann, M., Biehl, L. M., & Vehreschild, M. J. (2020). Fecal Microbiota Transfer. Deutsches Aerzteblatt Online, 31–38. https://doi.org/10.3238/arztebl.2020.0031

van Nood, E., Speelman, P., Nieuwdorp, M., & Keller, J. (2014). Fecal microbiota transplantation. Current Opinion in Gastroenterology, 30(1), 34–39. https://doi.org/10.1097/mog.0000000000000024

Wang, H., Cui, B., Li, Q., Ding, X., Li, P., Zhang, T., … Zhang, F. (2018). The Safety of Fecal Microbiota Transplantation for Crohn's Disease: Findings from A Long-Term Study. Advances in Therapy, 35(11), 1935–1944. https://doi.org/10.1007/s12325-018-0800-3

Wilson, B. C., Vatanen, T., Cutfield, W. S., & O'Sullivan, J. M. (2019). The Super-Donor Phenomenon in Fecal Microbiota Transplantation. Frontiers in Cellular and Infection Microbiology, 9. https://doi.org/10.3389/fcimb.2019.00002

Zeitz, J., Bissig, M., Barthel, C., Biedermann, L., Scharl, S., Pohl, D., … Scharl, M. (2017). Patients' views on fecal microbiota transplantation. European Journal of Gastroenterology & Hepatology, 29(3), 322–330. https://doi.org/10.1097/meg.0000000000000783

Zhang, T., Long, C., Cui, B., Buch, H., Wen, Q., Li, Q., … Zhang, F.

(2020). Colonic transendoscopic tube-delivered enteral therapy (with video): a prospective study. BMC Gastroenterology, 20(1). https://doi.org/10.1186/s12876-020-01285-0

Chapter 10.

Brandt L. J. (2012). Fecal transplantation for the treatment of Clostridium difficile infection. Gastroenterology & hepatology, 8(3), 191–194.

de Groot, P. F., Frissen, M. N., de Clercq, N. C., & Nieuwdorp, M. (2017). Fecal microbiota transplantation in metabolic syndrome: History, present and future. Gut Microbes, 8(3), 253–267. https://doi.org/10.1080/19490976.2017.1293224

Douglas, M. (1966). Purity and Danger. Routledge.

Ginkel, R., Strating, A., & van Ginkel, R. (2007). Wildness & Sensation. Transaction Pub.

Kahn, S. A., Gorawara-Bhat, R., & Rubin, D. T. (2012). Fecal bacteriotherapy for ulcerative colitis: Patients are ready, are we? Inflammatory Bowel Diseases, 18(4), 676–684. https://doi.org/10.1002/ibd.21775

Klasco, R. (2019, February 20). What Is a Fecal Transplant, and Why Would I Want One? The New York Times. https://www.nytimes.com/2019/01/18/well/live/what-is-a-fecal-transplant-and-why-would-i-want-one.html

Ma, Y., Yang, J., Cui, B., Xu, H., Xiao, C., & Zhang, F. (2017). How Chinese clinicians face ethical and social challenges in fecal microbiota transplantation: a questionnaire study. BMC Medical Ethics, 18(1). https://doi.org/10.1186/s12910-017-0200-2

McLeod, C., Nerlich, B., & Jaspal, R. (2019). Fecal microbiota transplants: emerging social representations in the English-language print media. New Genetics and Society, 38(3), 331–351. https://doi.org/10.1080/14636778.2019.1637721

Moossavi, S., Bishehsari, F., Ansari, R., Vahedi, H., Nasseri-Moghaddam, S., Merat, S., Sobhani, I., Keshavarzian, A., &

Malekzadeh, R. (2015). Minimum Requirements for Reporting Fecal Microbiota Transplant Trial. Middle East journal of digestive diseases, 7, 177–180.

Orduna, P., Lopez, S. Y., Schmulson, M., Arredondo, R., Ponce de Leon, S., & Lopez-Vidal, Y. (2015). A Survey Using the Social Networks Revealed Poor Knowledge on Fecal Microbiota Transplantation. Journal of Neurogastroenterology and Motility, 21(2), 294–295. https://doi.org/10.5056/jnm14146

The scoop on poop. (2020, June 23). St. Joseph's Health Care London. https://www.sjhc.london.on.ca/news-and-media/our-stories/scoop-poop

Zhang, F., Luo, W., Shi, Y., Fan, Z., & Ji, G. (2012). Should We Standardize the 1,700-Year-Old Fecal Microbiota Transplantation? American Journal of Gastroenterology, 107(11), 1755. https://doi.org/10.1038/ajg.2012.251

Chapter 11.

Borody, T., Leis, S., Campbell, J., Torres, M., & Nowak, A. (2021). Fecal Microbiota Transplantation (FMT) in Multiple in Multiple Sclerosis (MS). American Journal of Gastroenterology, 106(352). https://journals.lww.com/ajg/fulltext/2011/10002/fecal_microbiota_transplantation__fmt__in_multiple.942.aspx

Choi, H. H., & Cho, Y.-S. (2016). Fecal Microbiota Transplantation: Current Applications, Effectiveness, and Future Perspectives. Clinical Endoscopy, 49(3), 257–265. https://doi.org/10.5946/ce.2015.117

Khan, R., Petersen, F. C., & Shekhar, S. (2019). Commensal Bacteria: An Emerging Player in Defense Against Respiratory Pathogens. Frontiers in Immunology, 10(1203). https://doi.org/10.3389/fimmu.2019.01203

MayoClinic. (2021). Ulcerative colitis - Symptoms and causes. Mayo Clinic; https://www.mayoclinic.org/diseases-conditions/ulcerative-colitis/symptoms-causes/syc-20353326

Narula, N., Kassam, Z., Yuan, Y., Colombel, J.-F., Ponsioen, C., Reinisch, W., & Moayyedi, P. (2017). Systematic Review and Meta-analysis. Inflammatory Bowel Diseases, 23(10), 1702–1709. https://doi. org/10.1097/mib.0000000000001228

Reinink, A. R., Limsrivilai, J., Reutemann, B. A., Feierabend, T., Briggs, E., Rao, K., & Higgins, P. D. R. (2017). Differentiating Clostridium difficile Colitis from Clostridium difficile Colonization in Ulcerative Colitis: A Role for Procalcitonin?. Digestion, 96(4), 207–212. https://doi.org/10.1159/000481133

Schepici, G., Silvestro, S., Bramanti, P., & Mazzon, E. (2019). The Gut Microbiota in Multiple Sclerosis: An Overview of Clinical Trials. Cell Transplantation, 28(12), 1507–1527. https://doi. org/10.1177/0963689719873890

Suskind, D. L., Brittnacher, M. J., Wahbeh, G., Shaffer, M. L., Hayden, H. S., Qin, X., Singh, N., Damman, C. J., Hager, K. R., Nielson, H., & Miller, S. I. (2015). Fecal Microbial Transplant Effect on Clinical Outcomes and Fecal Microbiome in Active Crohn's Disease. Inflammatory Bowel Diseases, 21(3), 556–563. https://doi. org/10.1097/mib.0000000000000307

Wilson, B. C., Vatanen, T., Cutfield, W. S., & O'Sullivan, J. M. (2019). The Super-Donor Phenomenon in Fecal Microbiota Transplantation. Frontiers in Cellular and Infection Microbiology, 9(2). https://doi. org/10.3389/fcimb.2019.00002

Zeng, J., Peng, L., Zheng, W., Huang, F., Zhang, N., Wu, D., & Yang, Y. (2020). Fecal microbiota transplantation for rheumatoid arthritis: A case report. Clinical Case Reports, 9(2), 906–909. https://doi. org/10.1002/ccr3.3677

www.ingramcontent.com/pod-product-compliance
Lightning Source LLC
Chambersburg PA
CBHW021825190326
41518CB00007B/744